T0140552

An Introduction to Ultra-Fast Silicon Detectors

Series in Sensors
Series Editors: Barry Jones and Haiying Huang

Other recent books in the series:

An Introduction to Ultra-Fast Silicon Detectors

Marco Ferrero, Roberta Arcidiacono,
Marco Mandurrino, Valentina Sola,
Nicolò Cartiglia

CRC Press
Taylor & Francis Group
Boca Raton London New York

CRC Press is an imprint of the
Taylor & Francis Group, an **informa** business

First edition published 2021
by CRC Press
2 Park Square, Milton Park, Abingdon, Oxon, OX14 4RN

and by CRC Press
6000 Broken Sound Parkway NW, Suite 300, Boca Raton, FL 33487-2742

Library of Congress Cataloging-in-Publication Data

Names: Ferrero, Marco, author.
Title: Ultra-fast silicon detectors : design, tests, and performances /
Marco Ferrero, Roberta Arcidiacono, Marco Mandurrino, Valentina Sola,
Nicolo Cartiglia.
Description: First edition. | Boca Raton : CRC Press, 2021. | Series:
Series in sensors | Includes bibliographical references and index.
Identifiers: LCCN 2021020348 | ISBN 9780367646295 (hbk) | ISBN
9780367675936 (pbk) | ISBN 9781003131946 (ebk)
Subjects: LCSH: Semiconductor nuclear counters. | Very high speed
integrated circuits. | Particle tracking velocimetry. | Silicon diodes.
Classification: LCC TK9180 .U48 2021 | DDC 539.7/7--dc23
LC record available at https://lccn.loc.gov/2021020348

ISBN: 978-0-367-64629-5 (hbk)
ISBN: 978-0-367-67593-6 (pbk)
ISBN: 978-1-003-13194-6 (ebk)

Typeset in Nimbus
by KnowledgeWorks Global Ltd.

Contents

Preface

Silicon sensors have been one of the key detectors at the core of almost every high-energy physics experiment. In the past 30 years, silicon tracking systems have evolved from having a handful of electronic channels to the many millions present in the current detectors. In the past five years, the introduction of the low-gain avalanche diode design has opened up the possibility of using silicon sensors as precise time-tagging detectors. Due to this innovation, silicon sensors have gone from being known for their poor temporal resolution to be the detector of choice in applications that require very precise time tagging. This design innovation has produced a radical change in the design of future silicon particle trackers and allowed the introduction of 4D-tracking: the capability of tracking particles in space and time. This book describes our current understanding of how silicon sensors for 4D-tracking should be designed, the experimental techniques to test them in the laboratory, and a review of the most important results.

Acknowledgements

Our warmest thanks go to the colleagues that collaborated with us in the development of the Ultra-Fast Silicon Detectors. Special thanks to H. Sadrozinski, A. Seiden, and the whole team at our sister laboratory in Santa Cruz, to M. Boscardin, G.Paternoster, and the FBK UFSD team, and G.-F. Dalla Betta and L. Pancheri of Trento University. This book would not have been completed without the continuous help from the students of our group, led by M. Tornago and F. Siviero. They spent endless weeks in our laboratory testing and re-testing sensors, and endured long discussions without falling asleep. Their ingenuity is at the base of the results presented here. Finally, we want to thank the members of the CERN RD50 group for the very useful and productive collaborations.

The research program presented here would not have been possible without the generous financial contribution of the European Research Council via its ERC grant program.

1 Operating Principles of Silicon Sensors

This chapter presents a short introduction to the working principles of a silicon sensor, the signal formation, and the effects of radiation damage. The material is meant to highlight the basic notions that are used in the following chapters. The reader can find comprehensive reviews in several excellent recent publications. A clear and detailed explanation of the principles of operation of semiconductor sensors, with a specific emphasis on silicon, is presented in [97]. The evolution of silicon detectors is reported in [86], while an up-to-date publication of the radiation effects on the silicon bulk is presented in [102]. The latest trends in the construction of particle trackers are presented in [85]. Finally, an excellent reference is the *Review of Particle Physics* by the Data Particle Group [67].

A schematic cross section of a silicon sensor traversed by an ionizing particle is shown in Fig. 1.1, for a *p*-doped silicon sensor. An external bias voltage polarizes the *pn* junction inversely, creating a large depleted volume. The bias needed to fully deplete a silicon sensor, V_{FD}, is given by:

$$V_{FD} = \frac{q N_{A,eff} d^2}{2\varepsilon_{Si}}, \tag{1.1}$$

where q is the elementary charge, $N_{A,eff}$ the effective acceptor density, d the sensor active thickness, and ε_{Si} the silicon permittivity.

When a charged particle crosses the sensor, it creates along its path electron-hole (*e-h*) pairs, whose number depends on the particle type, energy, and sensor thickness. Under the influence of the electric field, the electrons drift towards the n^{++} implant while the holes towards the p^{++} implant, inducing a current signal on both electrodes. This current signal begins when the *e-h* pairs start moving and ends when the last charge carrier is collected at the electrodes.

1.1 ENERGY DEPOSITION IN SILICON

A charged particle traversing a silicon bulk interacts electromagnetically with the electrons of the atoms, losing energy gradually, and causing two different processes: (i) atomic excitation, displacing electrons to higher atomic orbitals, (ii) ionization, producing electron-ion pairs. The average energy loss by a particle per unit length is called *stopping power*, and it is given with good approximation by the Bethe-Bloch formula [2]:

$$-\frac{dE}{dx} = 4\pi N_A r_e^2 m_e c^2 \rho \frac{Z}{A} \frac{z^2}{\beta^2} \left[\frac{1}{2} \ln \frac{2 m_e c^2 \beta^2 \gamma^2 T_{max}}{I^2} - \beta^2 - \frac{\delta}{2} \right], \tag{1.2}$$

1

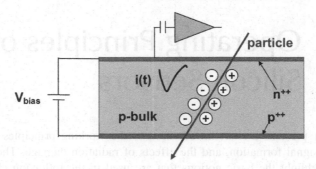

Figure 1.1 Schematic cross section of an n-in-p silicon sensor pad, traversed by an ionizing particle.

where E is the kinetic energy of the impinging particle of charge z, with moving velocity $\beta = v/c$ and Lorentz factor $\gamma = 1/\sqrt{1-\beta^2}$. I is the mean excitation energy of the target material, characterized by density ρ, and atomic and mass number Z and A. T_{max} is the maximum kinetic energy that could be transferred to a free electron in a single collision. N_A, r_e, and m_e are the Avogadro Number, the classic radius of the electron, and the mass of the electron, respectively, and δ is the high-energy corrective term for density. According to the Bethe-Bloch formula, the ionization loss is proportional to the electron density in the medium $\rho Z N_A/A$, to the particle charge squared, and it strongly depends on the incident particle velocity; at low momenta, the energy loss decreases proportionally to $1/\beta^2$. Figure 1.2 shows the energy loss for pions in silicon as a function of the pions momentum. The plot shows the energy deposition in an infinite silicon slab standard and in a thin 300 µm-thick silicon sensor (restricted). In thin layers, the deposited energy is lower because a fraction of the lost energy is carried off by energetic knock-on electrons [84]. A particle whose energy loss is at the minimum of the Bethe-Bloch function is called a Minimum Ionizing Particle (MIP).

The energy lost by a particle follows the Landau distribution, which is an asymmetric distribution with a not negligible tail at high energies due to delta rays [101]. Delta rays appear when a particle loses a large amount of its energy during a single interaction, and the electrons produced have enough energy to ionize other atoms. Due to this asymmetry, the mean value of the distribution does not match the most probable value (MPV), which is 30% lower. The energy needed to produce a single e-h pair in silicon is 3.6 eV, about three times the band gap since a large fraction of the energy is lost in lattice oscillations.

Both the MPV and the width of the Landau distribution are a function of the sensor thickness. This aspect is particularly relevant for the studies presented in the following chapters as the sensors under investigation are relatively thin, about 50 µm. The following two expressions, taken from [101], describe the Landau distribution

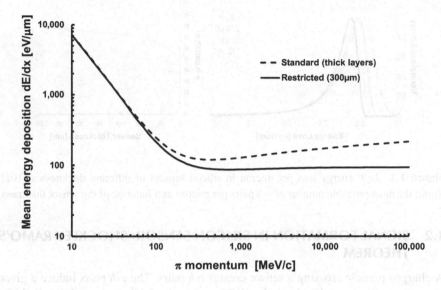

Figure 1.2 The mean energy deposition in silicon (Bethe-Bloch formula) as a function of pion momentum. The restricted curve shows the mean energy loss in a 300 μm-thick silicon sensor [84].

MPV and width as a function of the sensor thickness d:

$$MPV = 0.027 \times ln(d) + 0.126 \text{ [keV]} \tag{1.3}$$

$$width = 0.31 \times d^{0.81} \text{ [keV]}. \tag{1.4}$$

The evolution of the Landau distribution for increasing sensor thicknesses is shown on the *left* side of Fig. 1.3, while the *right* side shows the MPV of e-h pairs as a function of the sensor thickness. Thin sensors have smaller signals because their depletion region is narrower and because the mean number of e-h pairs per micron is smaller.

1.1.1 α PARTICLES

Laboratories often use radioactive sources, such as the americium-241, to generate α particles. Since the α particles are heavy, their stopping power is very large. For this reason, α particles lose all their energy within the first few microns after entering the silicon sensor. They are the ideal tool if a large and localized creation of e-h pairs is needed.

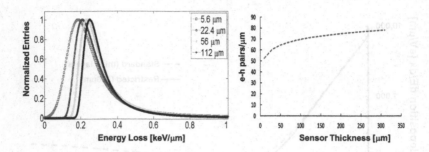

Figure 1.3 *Left*: energy loss per micron in silicon sensors of different thicknesses [101]. *Right*: the most probable number of *e-h* pairs per micron as a function of the sensor thickness.

1.2 SIGNAL FORMATION IN SILICON SENSOR: SHOCKLEY-RAMO'S THEOREM

A charged particle crossing a sensor creates *e-h* pairs. The *e-h* pairs induce a given amount of charge on the electrodes. When an external bias voltage is applied, the *e-h* pairs begin to move: since the amount of induced charge changes with their position, their motion creates a current. The signal induced on a given electrode is calculated using the Shockley-Ramo's theorem [112, 119]. The Shockley-Ramo's equation calculates the induced current $i_k(t)$ on the electrode k by a charge q as the scalar product of the drift velocity $\vec{v}$ with the weighting field $\vec{\mathscr{E}}_w$:

$$i_k(t) = -q\vec{v} \cdot \vec{\mathscr{E}}_w. \tag{1.5}$$

The weighting field $\mathscr{E}_w$ describes the coupling between the charge q and the k-th electrode. The weighting field corresponds numerically (even though it has dimensions $[L^{-1}]$) to an electric field calculated setting at 1 V the read-out electrode and at 0 V all other electrodes. For this reason, the weighting field has the same dependence upon the geometry of the electrodes of an electric field: it decreases with distance d as $1/d^2$ if the electrode is a point, as $1/d$ if it is a line, and it is constant between two large electrodes.

Consider a single *e-h* pair inside the bulk of a sensor. The drifts of the *e* approaching the cathode and that of the *h* moving away from it induce on this electrode two currents with the same sign. The integral of these two currents is equal to the unity of charge q:

$$\int (i_e(t) + i_h(t))dt = q. \tag{1.6}$$

Even though the integral is always equal to q, the relative contributions of $i_e(t)$ and $i_h(t)$ to the signal depend upon the geometry of the electrodes and the applied electric field. In a straightforward configuration, the sensor has the geometry of a parallel plate capacitor, with the anode and cathode representing the two plates. In this case, the weighing field is constant, and it is equal to $\mathscr{E}_w = 1/d$. With a constant weighing field, a charge's induced signal depends uniquely upon its velocity and not upon its position. An opposite example is a strip sensor with a large pitch and a narrow strip implant. In this case, the weighing field at a distance x from the strip is $\mathscr{E}_w = 1/x$ and, consequently, the signal is generated only when the charge carriers are very near the electrode. In this configuration, assuming a read-out connected to the cathode, only the electrons are generating a sizable signal. With this geometry, if a particle creates e-h pairs between two strips, the signal will not be seen until the electrons are near the cathode, while the holes do not contribute significantly to the signal.

Figure 1.4 shows two examples of the electrons, holes, and total simulated currents for a 100 µm-thick sensor with a single large strip (*left*) or many narrow strips (*right*). For the sensor with narrow strips, the electrons current is larger as the weighting field is concentrated near the cathode.

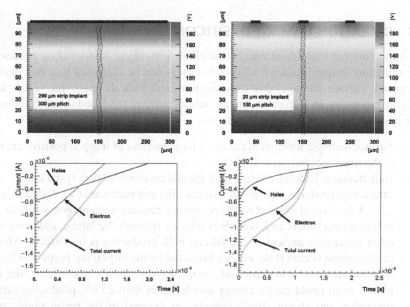

Figure 1.4 *Top*: 2D map of the electric potential generated in a 100 µm-thick sensor volume (cross-cut, electrodes on top), biased at 200 V, for two different configurations: a single large strip (*left*) or three narrow strips (*right*). *Bottom*: electrons, holes, and total currents as simulated in the above mentioned devices.

The second ingredient of Eq. (1.5) is the charge carriers drift velocity, shown in Fig. 1.5. The effect of a higher velocity is to make the signal shorter and sharper. The electrons drift velocity saturates at room temperature when the electric field is about $\mathscr{E} = 30$ kV/cm while the holes drift velocity does not saturate; it increases for values of the field up to $\mathscr{E} = 100$ kV/cm.

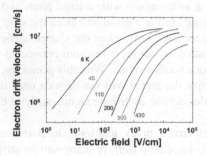
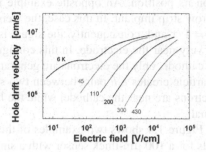

Figure 1.5 Electrons (*left*) and holes (*right*) drift velocities as a function of the electric field [92] at several temperatures.

1.3   RADIATION DAMAGE IN SILICON SENSORS

An impinging particle looses energy via either ionizing or non-ionizing processes. The radiation damage induced in silicon sensors can be classified into two distinct categories: surface damage, due to ionization, and bulk damage, due to the non-ionizing processes. In both cases, the amount of damage produced depends upon the type and energy of the impinging particles.

1. Surface damage: irradiation creates a large number of trapped positive charges at the interfaces between Si-SiO_2.
2. Bulk damage: hadrons interact with the silicon atoms of the crystal lattice. This interaction produces silicon interstitials (Si_i) and vacancies (V), called Frenkel pair. A fraction of Frenkel pairs recombine, causing no damages, while the remaining interstitials and vacancies migrate through the lattice and react with other impurities present in the silicon bulk producing point defects. Atomic displacement occurs if the energy imparted by the impinging particle is higher than the displacement threshold energy E_d (~ 25 eV). Sometimes, the displaced atom could gain a energy much higher than 25 eV, producing further ionization and atomic displacements. At the end of the recoil range, non-ionizing reactions prevail, producing dense agglomeration of defects called clusters.

The energy lost which does not go into ionization is called *Non Ionizing Energy Loss (NIEL)* [134], and is normally reported in keV·cm²/g. A fraction of this energy produces lattice excitation, while the other fraction is responsible for bulk damages. In order to compare the effects induced by different particles (in type or energy), the hypothesis that the radiation damage scales with their NIEL factors was introduced (*NIEL scaling hypothesis*). The reference NIEL value has been chosen to be that of 1 MeV neutrons.

Using the NIEL hypothesis, the studies carried out on irradiated devices are compared to each other by converting each fluence to their equivalent 1 MeV neutron fluence. For this reason, irradiation fluences are often reported in numbers of

1 MeV n_{eq}/cm^2. Figure 1.6 shows the NIEL factor in silicon for neutrons, protons, pions, and electrons normalized to the NIEL value of 1-MeV neutrons. Given their

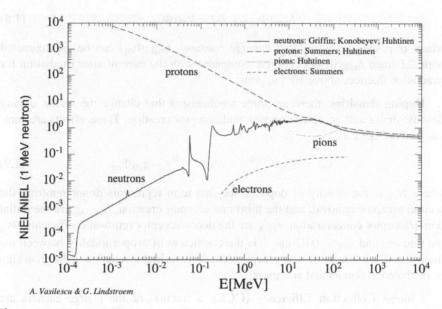

A. Vasilescu & G. Lindstroem

Figure 1.6 Relative NIEL factor in silicon for neutrons, protons, pions, and electrons, as a function of the particle energy [113]. The NIEL factor of 1-MeV neutrons is used as normalization value.

very high dE/dx, low energy protons produce a much larger displacement damage than neutrons. Above about 50 MeV, the NIEL values of neutrons and protons become very similar.

1.3.1 IMPACT OF DEFECTS ON THE PROPERTIES OF SILICON SENSORS

Radiation creates silicon interstitials, vacancies, and clusters which, at room temperature, migrate through the silicon lattice, react with impurities, and give rise to other defects. In this subsection, the impact of these defects on the properties of silicon sensors will be recalled, with special attention to effects relevant to Ultra-Fast Silicon Detectors (UFSDs). For an up-to-date review of displacement damage in silicon sensors see [102].

Leakage current (I_{bulk}): defects close to the middle of the band gap are generator centers of electron-hole pairs, hence responsible for the increase of the current. The leakage current is proportional to the depleted volume V_d, the intrinsic carrier density n_i, and the inverse of the generation lifetime (τ_g):

$$I_{bulk} = V_d q_o \frac{n_i}{\tau_g}, \qquad (1.7)$$

where q_o is the electron charge. In irradiated sensors, the increase in current ($\triangle I$) is proportional to the 1-MeV equivalent fluence Φ_{eq}, Eq. (1.8):

$$I_{bulk}(\Phi_{eq}) \sim \triangle I = V_d \alpha \Phi_{eq}, \tag{1.8}$$

where α is the *current-related damage constant*. $I_{bulk}(\Phi_{eq})$ can be approximated with $\triangle I$ since $I_{bulk}(0)$ is negligible compared with the current after irradiation for irradiation fluences above 10^{12} n_{eq}/cm^2.

Doping densities: there are three mechanisms that change the sensor doping density: donor and acceptor removal, and acceptor creation. These effects are summarized in Eq. (1.9):

$$N_{eff} = N_{D_0} e^{-c_D \Phi_{eq}} - N_{A_0} e^{-c_A \Phi_{eq}} - g_{eff} \Phi_{eq}, \tag{1.9}$$

where N_{eff} is the density of dopants, the first term represents donor removal, the second acceptor removal, and the third one acceptor creation. N_{D_0/A_0} are the initial donor/acceptor concentration, $c_{D/A}$ are the donor/acceptor removal coefficients, Φ_{eq} the fluence, and $g_{eff} = 0.02$ cm^{-1} is the coefficient of proportionality between the fluence and the density of new acceptor-like defects. In the equation, a different sign is attributed to donors and acceptors[1].

Charge Collection Efficiency (CCE): a fraction of the charge carriers are trapped by deep defects during their drift. This process, called trapping, decreases the charge collection efficiency. The probability of being trapped depends on the drift's duration and not upon its length: more charge is collected at high drift velocities. The current signal $i(t)$ in an irradiated sensor follows an exponential decrease as a function of time:

$$i(t) = i(t)_{\Phi_{eq}=0} \times e^{-\frac{t}{\tau_{eff}}}. \tag{1.10}$$

The time constant τ_{eff} is given by:

$$\frac{1}{\tau_{eff}} = \Phi_{eq} \beta_{e/h}, \tag{1.11}$$

where $\beta_{e/h}$ [cm^2/ns] are proportionality factors called *effective trapping damage constant*. τ_{eff} decreases as a function of the irradiation fluence and the $\beta_{e/h}$ parameter. If the de-trapping time is longer than the shaping time of the read-out electronics, the trapping will result in a decrease of the CCE (i.e., the ratio of the number of collected charges over the number of generated ones). Equation (1.11) predicts a linear decrease of τ_{eff} with fluence, however, as reported in [41], this is not accurate at high fluences, where the correct expression is[2]:

$$\tau_{eff} = 540 \cdot \Phi_{eq}^{-0.62} \text{ [ps]}, \tag{1.12}$$

where Φ_{eq} is in unit of 10^{15} n_{eq}/cm^2.

---

[1]Even though the Greek letter ρ is often used to indicate densities, it is an established tradition to use the symbol $N_{A,D}$ for the acceptor and donor densities.

[2]Note: the paper [41], due to a clerical error, reports the constant to be 54.

Decreased carriers mobility: The electrons and holes mobilities decrease with fluence as the number of scattering centers increases [41]. Indicating with $\mu_{0,\text{sum}}$ the sum of the electrons and holes zero-field mobilities, then the dependence of $\mu_{0,\text{sum}}$ upon the fluence at a fixed temperature of -20°C can be expressed as:

$$\mu_{0,\text{sum}} = 3500 \text{ cm}^2/\text{Vs} \cdot \left(\frac{\Phi_{\text{eq}}}{10^{15}\text{n}_{\text{eq}}/\text{cm}^2}\right)^{-0.46}, \tag{1.13}$$

which can be used for $\Phi_{\text{eq}} > 5 \cdot 10^{15}$ $\text{n}_{\text{eq}}/\text{cm}^2$. A lower mobility implies a lower charge carries kinetic energy and, ultimately, an impossibility to reach controlled multiplication. If this will be confirmed by experiments, then the mobility reduction with fluence would limit the use of UFSDs in environments with very high radiation levels.

1.3.2 ACCEPTOR REMOVAL

As explained in Section 1.3.1, the process of acceptor removal consists of the progressive reduction of the number of active acceptor atoms. This effect is particularly damaging in UFSDs as it deactivates the acceptor-doped implant (the so-called *gain implant*) responsible for the controlled internal multiplication process. Given its importance, this topic is presented in detail in the following.

The microscopic origin of acceptor removal [35] is not fully understood, and it is still under investigation [103]. An important consideration is that Secondary Ion Mass Spectrometry (SIMS) measurements have demonstrated that boron atoms have not been removed, they are simply not electrically active. SIMS have the capability of measuring the density of boron, substitutional and interstitial, as a function of depth, so they can assess the differences between the gain implants of an unirradiated UFSD with that of a highly irradiated UFSD, where the gain mechanism is almost absent. For this study, SIMS has been made on two twins UFSDs, the first (M83) unirradiated while the second (M80) irradiated with a fluence of $1 \cdot 10^{16}$ $\text{n}_{\text{eq}}/\text{cm}^2$. In this second UFSD, the gain mechanism has been completely canceled by radiation damage. The SIMS results show identical boron profiles for the two samples, indicating that the disappearance of the gain does not correspond to the removal of the boron atoms, only to their inactivation, Fig. 1.7. A possible explanation of the acceptor removal mechanism is based on the ion-acceptor complexes formation with irradiation. Irradiation moves silicon atoms outside the lattice creating silicon interstitial. Si_i interacts with substitutional impurities such as B_s via the so called Watkins replacement mechanism:

$$\text{Si}_i + \text{B}_s \rightarrow \text{B}_i. \tag{1.14}$$

In this interaction, the impurities become interstitial (B_i); B_i interacts with other interstitial impurities such as O_i forming B_iO_i complexes:

$$\text{B}_i + \text{O}_i \rightarrow \text{B}_i\text{O}_i. \tag{1.15}$$

In total, the radiation removes an acceptor level (shallow level) from the band gap, introducing a donor level.

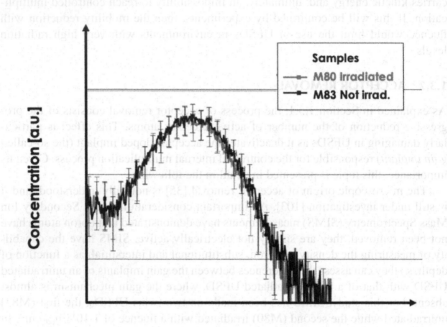

Figure 1.7 SIMS measurements of the density of boron atoms forming the gain implant as a function of the depth, for unirradiated (M83), and heavily irradiated (M80, fluence about $1 \cdot 10^{16}$ n_{eq}/cm^2) UFSDs. Although the gain implant of the M80 sample is almost completely deactivated, the two doping profiles are identical [48].

Empirically, acceptor removal can be modelled as follow [48]:

$$N_A(\Phi_{eq}) = N_A(0) \cdot e^{-c_A \Phi_{eq}} = N_A(0) \cdot e^{-\Phi_{eq}/\Phi_{o\,eq}}, \qquad (1.16)$$

where Φ_{eq} is the irradiation fluence [cm^{-2}], $N_A(0)$ ($N_A(\Phi_{eq})$) the initial (after a fluence Φ_{eq}) acceptor density [cm^{-3}], and c_A [cm^2] is a constant that depends on the acceptor concentration $N_A(0)$ and on the type of irradiation. $\Phi_{o\,eq} = 1/c_A$ is the value of fluence that reduces the initial doping density $N_A(0)$ to $1/e$ of its initial value. $\Phi_{o\,eq}$ can be written as:

$$\Phi_{o\,eq} = \frac{0.63 \cdot N_A(0)}{\rho_{Si} \cdot k_{cap} \cdot N_{Int} \cdot \sigma_{Si}}, \qquad (1.17)$$

where $0.63 \cdot N_A(0)$ is the density of removed acceptor, $\rho_{Si} = 5 \cdot 10^{22}$ cm^{-3} is the silicon atomic density, N_{Int} the number of defects created in the interaction, k_{cap} the probability for a defect to capture an acceptor, and σ_{Si} the cross-section between radiation and silicon. The capture coefficient k_{cap} depends upon the doping used for the gain implant and the presence of additional impurities such as carbon or oxygen.

The product $k_{cap} \cdot N_{Int}$ is the maximum number of deactivated acceptor atoms per incident particle. This number is reached only when each defect is near an acceptor, so at high acceptor densities. At low acceptor densities, defects and acceptors are too far apart and there is no interaction. This effect is quantified by the proximity function D:

$$D = \frac{k_{cap} \cdot N_{Int}}{1 + (\frac{N_{Ao}}{N_A(0)})^{2/3}}, \qquad (1.18)$$

where $N_A(0)$ is the acceptor density, and N_{Ao} is a fit parameter to be obtained from the data. Even though the number of removed acceptors is higher for higher acceptor densities, the fraction of removed acceptors is larger at lower densities, *left* side of Fig. 1.8. For this reason, acceptor removal has a more substantial impact at low acceptor densities.

Combining Eq. (1.17) with Eq. (1.18) normalized to $k_{cap}N_{Int}$, the value of the acceptor removal coefficient $c_A = 1/\Phi_{o\,eq}$ can be expressed as:

$$c_A = \frac{\rho_{Si} \cdot \sigma_{Si} \cdot k_{cap} \cdot N_{Int}}{0.63 \cdot N_A(0)} \cdot \frac{1}{1 + (\frac{N_{Ao}}{N_A(0)})^{2/3}}. \qquad (1.19)$$

Figure 1.9 shows the measured acceptor removal coefficient c_A as a function of the acceptor density $N_A(0)$, for several silicon sensors. It also shows the parametrization in Eq. (1.19) fitted to the experimental data.

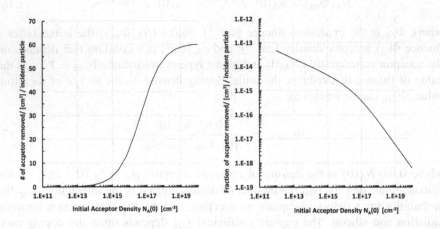

Figure 1.8 *Left:* number of acceptors per cm^3 removed per incident particle, as a function of the acceptor density. At the highest density, ~ 60 acceptors/cm^3 are removed per incident particle. *Right:* fraction of acceptors per cm^3 removed per incident particle as a function of the acceptor density.

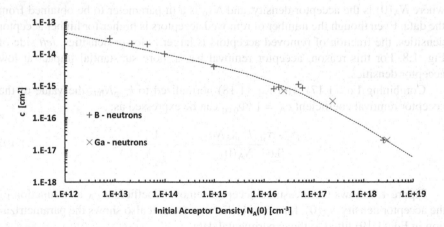

Figure 1.9 Acceptor removal coefficient c_A as a function of the initial acceptor density $N_A(0)$ [48].

2 Ultra-Fast Silicon Detectors

Ultra-Fast Silicon Detectors are an innovative type of pixelated silicon sensor able to measure concurrently the location and the time of a hit with very good accuracy. In particular, the objective of a UFSD device is to achieve a temporal resolution of ~ 30 ps and a spatial resolution of ~ 10 μm. This is obtained by combining a moderate intrinsic gain (~ 20) with a fine segmentation of the sensor.

The initial part of this chapter presents the Low-Gain Avalanche Diode (LGAD) technology and how this technology has been optimized in the UFSD design to achieve excellent temporal resolution. The second part describes the UFSD segmentation technology required for the design of large-area sensors; the last part of the chapter discusses the effects of radiation damage in UFSD and the possible technological solutions to improve their radiation resistance. Further information can be found in [64, 60, 62, 65, 118, 75].

2.1 LOW-GAIN AVALANCHE DIODE TECHNOLOGY

In silicon sensors, charge multiplication happens when charge carriers drift in a region with an electric field ($\mathcal{E}$) greater than about 300 kV/cm [98]. Under this condition, the electrons (and to less extent the holes) acquire enough kinetic energy to produce, by impact ionization, additional e-h pairs.

The gain of the avalanche process is defined as the ratio between the total number of e-h pairs over the number of e-h pairs collected in the absence of multiplication ($G = N_{e,h}/N_{0;e,h}$). The avalanche process is commonly used in semiconductor sensors such as the Avalanche Photodiodes (APD) and Silicon Photomultipliers (SiPM) [23], with a gain of about 100 and 10,000, respectively.

In a standard silicon sensor, the so-called PIN diode[1], a high value of the electric field is obtained applying a high external bias voltage. This condition is prone to cause a device breakdown since it induces a very high field on the device periphery.

On the contrary, in the LGAD design [40], the electric field value of ~ 300 kV/cm is obtained in a very localized region by depleting an additional p^+-doped layer (boron or gallium) implanted near the pn junction. Schematics of an n-in-p PIN diode and of an LGAD are shown in Fig. 2.1. The additional p^+ layer, 0.5–1 μm wide implanted at a depth of about 0.5–2 μm, is characterized by an acceptor density of about $N_A \sim 10^{16}$ atoms/cm^3 that, when depleted, locally generates an electric field high enough to activate the avalanche process. In this design, the high field region is underneath the n^{++} implant and does not extend to the sensor periphery.

The LGAD design provides a moderate gain, of the order 10–30, and merges the best characteristics of standard silicon sensors (low noise, segmentation, low

---

[1]The term PIN diode refers to a structure composed of a sequence of p-doped/intrinsic/n-doped silicon. Standard silicon sensors are often called PIN diode, even though the bulk is not made of intrinsic silicon but is lightly doped.

leakage current, blindness to low energy photons) with those of APDs and SiPMs (large signals and good temporal resolution).

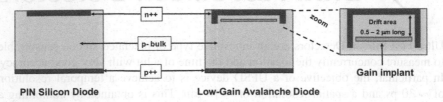

Figure 2.1 A schematic view of: (*left*) an *n*-in-*p* PIN diode; (*right*) a Low-Gain Avalanche Diode. The LGAD design is characterized by the presence of an additional p^+ implant underneath the *pn* junction.

2.1.1 CHARGE MULTIPLICATION

As already mentioned in Section 2.1, the avalanche mechanism starts when a charge carrier drifts in a region with a high electric field. This effect, often called impact ionization mechanism, is described by Eq. (2.1). The number $N_{e,h}(d)$ of *e-h* pairs generated by the avalanche has an exponential dependency on the impact ionization coefficient $\alpha_{n,p}$ and on the length d travelled inside the high electric field region:

$$N_{e,h}(d) = N_{e,h}(0)e^{\alpha_{n,p}d}. \tag{2.1}$$

The inverse of the ionization coefficient α is the mean free path between two subsequent scattering events producing secondary charges, $\lambda = 1/\alpha$. For a given electric field, this distance is shorter for electrons (λ_n) than for holes (λ_p). Therefore, it is possible to tune the electric field to values where only the electrons multiply. This possibility allows having a low multiplication factor since the avalanche never develops. Impact ionization occurs, on average, when a charge carrier travels for a distance long enough ($\sim \lambda$) to acquire a kinetic energy greater or equal to the lowest ionization energy E_i. E_i, using the conservation law of momentum and energy, can be estimated to be about $1.5E_g$, where E_g is the energy gap of the semiconductor (1.12 eV for silicon at 300 K).

A simplified avalanche multiplication model in silicon, similar to the model used in gases, is the Chynoweth model [7]:

$$\alpha_{n,p}(\mathscr{E}) = \frac{1}{\lambda_{n,p}(\mathscr{E})} = A_{n,p}e^{-\frac{B_{n,p}}{\mathscr{E}}}, \tag{2.2}$$

where $\alpha_{n,p}$ are the electron and hole ionization coefficients, $A_{n,p}$ the maximum number of *e-h* pairs that can be generated in presence of a very high electric field, $\mathscr{E}$ the electric field, and $B_{n,p}$ the coefficients derived from experimental fits. Numerical values of $A_{n,p}$, and $B_{n,p}$ can be found in [5, 52, 131]. The functional form of the $B_{n,p}$ coefficients is shown in Eq. (2.3):

$$B_{n,p}(T) = C_{n,p} + D_{n,p}T. \tag{2.3}$$

The dependence of the $B_{n,p}$ coefficients upon the temperature introduces a relationship between the ionization coefficients and temperature: the mean distance necessary to achieve multiplication is shorter at lower temperature (minor phonon population), yielding to an increase of gain at the same electric field. Impact ionization models, commonly used in numerical simulation [128], see also Section 3.4.1, are van Overstraeten-de Man [133], Massey [100], Okuto-Crowell [105], and the Bologna model [70]. Figure 2.2 shows the relationship between the mean distance and the electric field at two temperatures, 250 K and 300 K, for three models (Massey, van Overstraeten-de Man, and Okuto-Crowell).

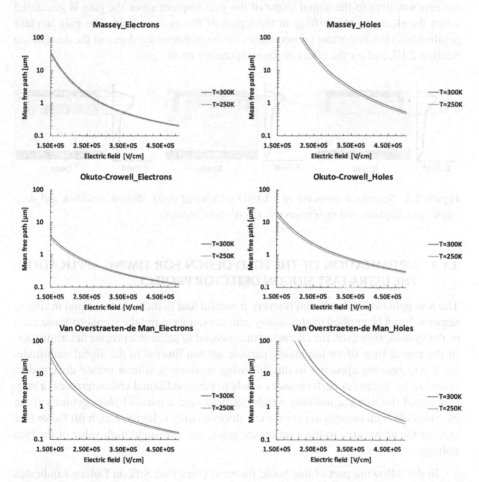

Figure 2.2 Mean free path λ_n for electrons (*right*) and holes (*left*) as a function of the electric field, for the three avalanche multiplication models: Massey, van Overstraeten-de Man, and Okuto-Crowell. The lighter (darker) line shows the mean free path at 250 K (300 K).

An important parameter in the design of an LGAD is the depth of the gain implant. Consider the three LGADs shown in Fig. 2.3: (i) one with a very broad implant, in contact with the n^{++} electrode, (ii) one with the gain implant separated but near the n^{++} electrode, and (iii) one with a deep gain implant. These three cases are referred to as the broad, shallow, and deep designs, respectively. If the maximum electric field were the same in the three designs, the gain would be more significant in the deep design since the number of mean free paths contained in the gain layer is higher. For this reason, to obtain the same gain in all designs, the field needs to decrease as a function of the implant depth. The broad design is the most delicate: most of the gain happens very near the *pn* junction, and small process variations can lead to very different gain values. On the contrary, the shallow and deep designs are less sensitive to the actual shape of the gain implant since the gain is generated when the electrons are drifting in the region of flat electric field. The gain implant position also has important consequences on the radiation hardness of the design, see Section 2.10, and on the effect of the temperature on the gain.

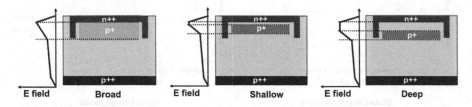

Figure 2.3 Schematic cross cut of a UFSD with broad (*left*), shallow (*middle*), and deep (*right*) gain implants and their respective electric field profiles.

2.1.2 OPTIMIZATION OF THE LGAD DESIGN FOR TIMING APPLICATION: THE ULTRA-FAST SILICON DETECTOR PROJECT

The low-gain avalanche design is a very powerful tool to increase the signal in silicon sensors. It can be applied to almost any geometry, improving the signal-to-noise ratio of the system. However, the characteristics needed to achieve a precise determination of the arrival time of an impinging particle are not limited to the signal amplitude. As it will become apparent in the following sections, a silicon sensor designed to optimize the temporal performances needs to have additional characteristics: a very fast signal (be thin), a uniform weighting field (use a parallel plate geometry), an electric field high enough to saturate the electrons drift velocity, a high fill factor (the ratio of the active area to the total sensor area), and withstand high value of the bias voltage.

In the following part of this book, the term Ultra-Fast Silicon Detector indicates LGADs whose design has been optimized for timing applications. It is worth noticing that in the literature, the terms LGAD and UFSD are often used as synonyms.

2.2   THE WEIGHTFIELD2 SIMULATION PROGRAM

The Weightfield2 simulation program[2] [26] has been developed with the specific aim of reaching a better understanding of the properties of UFSDs. The program has been extensively validated comparing its predictions to laboratory measurements, and it has been a valuable tool in the design and evaluation of UFSDs. The program has a graphical user interface (GUI), shown in Fig. 2.4, where the user enters the parameters of the simulation. The GUI is divided into different areas. The left side has five tabs showing the electric potential, the weighting potential, the generated currents, and electronics I and II. The central column controls the program flow and its output files, and hosts the selection of the signal source (α, MIP, laser, etc.), the irradiation level, and the presence of a magnetic field. The column on the right controls the sensor geometry, the presence of the gain mechanism, and simplified read-out models for a few typical amplifiers.

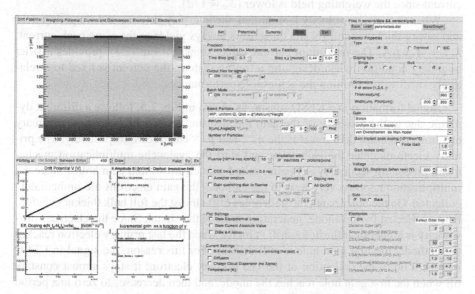

Figure 2.4   The graphical user interface of the Weightfield2 simulation program.

Throughout the following chapters of this book, WF2 simulations are used to illustrate the main UFSDs characteristics and provide guidance in understanding the experimental measurements. In the actual design of UFSDs, these simulations need to be complemented by those obtained with TCAD simulations, see Chapter 3.

2.3   UFSD SIGNAL FORMATION

To understand the characteristics of the signal generated in a UFSD, it is necessary to study first the mechanisms of signal formation in a PIN diode. The current in a PIN

---

[2]Freely available at http://l.infn.it/wf2

diode of active thickness d can be calculated using the Ramo-Shockley's theorem, shown in Eq. (1.5). Assuming a pixel size much larger than the sensor thickness, the weighting field is a constant, $\mathcal{E}_w \propto 1/d$. Assuming also an electric field high enough to saturate the drift velocity v_{sat}, the initial current can be written as:

$$I_{max} \propto Nq\frac{1}{d}v_{sat} = (n_{e,h}d)q\frac{1}{d}v_{sat} = n_{e,h}qv_{sat}, \qquad (2.4)$$

where N is the number of e-h pairs along the sensor thickness d, and q is the electric charge. N is given by the product between the sensor thickness d and the number of e-h pairs $n_{e,h}$ per unit length (assuming uniform ionization). Equation (2.4) shows a very interesting result: the maximum current in a PIN diode does not depend on the sensor thickness, only on the carriers velocity. In thicker sensors, more e-h pairs are ionized (N is proportional to d); however, each charge contributes less to the initial current since the weighting field is lower ($\mathcal{E}_w \propto 1/d$).

When the carriers velocity is saturated, the interplay between the weighting field and sensor thickness in PIN silicon sensors always yields to the same peak current, $I_{max} \sim 1.5 \ \mu A$. The maximum current value, shown in Eq. (2.4), provides a strict boundary condition when designing a timing system: the current is fixed to a relatively small value, limiting the achievable performances.

Signal formation in a UFSD sensor, as shown in Fig. 2.5, follows a different dynamic due to the presence of the gain mechanism. As in a PIN diode, the primary electrons and holes drift towards the n^{++} and p^{++} electrodes, respectively. The primary electrons enter the gain layer and start the avalanche multiplication mechanism, producing secondary e-h pairs, called gain electrons and gain holes. Since the multiplication happens very near to the cathode, the gain electrons are immediately collected. On the other hand, the gain holes drift almost the full bulk thickness before being collected by the anode, generating most of the signal. Since the electrons drift velocity is higher than the holes drift velocity, when the last primary electron reaches the cathode, the first gain hole is still drifting. For this reason, the signal in UFSD increases up to the collection of the last primary electron; it stays almost constant till when the first gain hole reaches the anode, and then decreases to zero in a period controlled by the holes drift velocity.

Since the weighting field is constant in the bulk, the drift of gain holes generates a large induced current till they reach the anode. In UFSDs, the gain holes current constitutes the largest contribution to the total current, see Fig. 2.5.

The time length of a signal in UFSD is longer than that in PIN diodes. In PIN diodes, the signal length is determined by the holes drift time while in UFSD by the sum of the electrons and holes drift times. The rise time is also very different: in a PIN diode it is almost instantaneous since it is the time it takes for the ionized e-h pairs to reach their drift velocity, while in UFSD the rise time is equal to the electrons drift time.

The current generated by the multiplication mechanism can be estimated from the number of electrons entering the gain layer in a time interval dt, assuming a drift velocity v_{sat}. The amount of these primary electrons is $n_{e,h}v_{sat}dt$ and they generate a

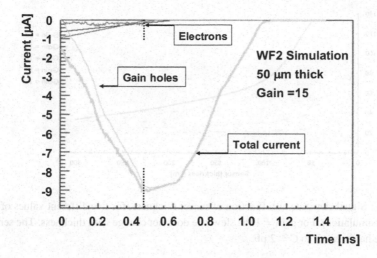

Figure 2.5 WF2 simulation of the total current, with the various contributions due to primary electrons/holes and gain electrons/holes, generated by a MIP traversing a 50 μm-thick UFSD. The current increases for the duration of the electron current.

number of e-h pairs equal to $dN_{\text{gain}} \propto n_{e,h}(v_{\text{sat}}dt)G$. Using the Ramo-Shockley's theorem, and assuming a parallel plate geometry ($\mathscr{E}_w = 1/d$), it is possible to calculate the current induced by these secondary charges:

$$dI_{\text{gain}} = qv_{\text{sat}}\frac{1}{d}dN_{\text{gain}} \propto \frac{G}{d}dt, \tag{2.5}$$

which leads to:

$$\frac{dI_{\text{gain}}}{dt} \sim \frac{dV}{dt} \propto \frac{G}{d}. \tag{2.6}$$

Equation (2.6) shows a key feature of UFSDs: the signal slew rate, dI/dt, is proportional to the ratio between the sensor gain and thickness (G/d). This implies that thin sensors with high gain have signals with fast slew rates and, therefore, are suitable to perform high-precision temporal measurements.

Using the WF2 program, the slew rate for UFSDs of different thicknesses and gains has been evaluated, see Fig. 2.6. In a 300 μm-thick UFSD, the slew rate increase with gain is limited by the large value of d: at gain 20 the slew rate is only twice that of a PIN sensor. On the other hand, in a 50 μm-thick UFSD at gain 20, the slew rate is more than six times higher.

Equivalently to the calculation shown at the beginning of this section for the PIN sensor, it is possible to calculate the maximum current in a UFSD. Since each primary electron generates G e-h pairs, I_{max} can be written as:

$$I_{\text{max}} \propto N_{\text{max}}q\frac{1}{d}v_{\text{sat}} = (n_{e,h}dG)q\frac{1}{d}v_{\text{sat}} = n_{e,h}Gqv_{\text{sat}}. \tag{2.7}$$

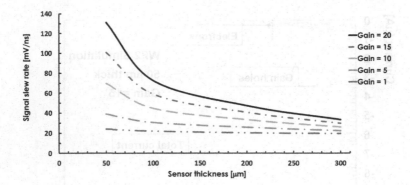

Figure 2.6 Signal slew rate as a function of sensor thickness, for five different values of the gain (WF2 simulation). For gain $= 1$ the slew rate does not change with thickness. The sensor capacitance has been set to $C = 2$ pF.

Equation (2.7) shows that, when the drift velocity is saturated, the signal peak current only depends on the gain and not on the sensor thickness. On the other hand, the sensor thickness determines the rise time of the signal. Figure 2.7 schematically shows the signal shapes for sensors with equal gain and different thicknesses.

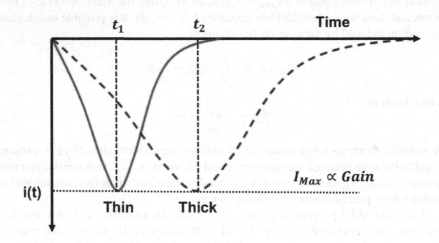

Figure 2.7 Signals generated in UFSD devices with the same gain and different active thicknesses.

2.4   UFSD NOISE SOURCES

In every silicon sensor, the flow of electrons over the *pn* junction generates a fluctuation of the current commonly known as shot noise. The shot noise is the root mean-square current fluctuation of the current I in a given bandwidth interval Δf:

$$\sigma_{shot} = \sqrt{2q(I_{surface} + I_{bulk})\Delta f},\qquad(2.8)$$

where q is the electron charge and $I_{surface}$, I_{bulk} the surface and bulk leakage currents, respectively.

In PIN diodes, the shot noise is fairly small, and it does not represent a dominant source of noise. On the contrary, in a sensor with high internal gain, the shot noise might become significant since the current is multiplied by the gain value. UFSDs typically operate at a relatively low gain (10–30), a condition where the shot noise usually is subleading with respect to the electronic noise. However, as shown in Section 2.9, after heavy irradiation, the value of the leakage current increases, and the shot noise can be the dominant contribution to the total noise.

In sensor with gain, a second effect, the so-called excess noise, contributes to make the shot noise term larger. The excess noise is an additional noise induced by the multiplication mechanism: each primary electron entering the gain layer generates a number of secondary charges that, on average, is equal to G. However, since the multiplication is a random process, each electron generates a number of secondary charges that is not exactly G. This variability increases the shot noise by a factor F, the so-called excess noise factor. F is a function of G and it is given by [31]:

$$F \sim G^x = Gk + (2 - \frac{1}{G})(1 - k),\qquad(2.9)$$

where x is called excess noise index and $k = \alpha_p/\alpha_n$ is the ratio between the impact ionization coefficients of holes and electrons.

For UFSDs, Eq. (2.8) needs therefore to incorporate the gain G and the excess noise factor F:

$$\sigma_{shot} = \sqrt{2q(I_{surface} + I_{bulk}G^2F)\Delta f}.\qquad(2.10)$$

In order to keep the shot noise small, the gain should be kept low, and the factor F must be reduced as much as possible. A way to minimize F is to use n-in-p UFSD: in this design, the avalanche is started by the electrons and the value of α_p can be kept small [60] since the holes do not contribute to the avalanche. Using Eq. (2.10), the signal-to-noise ratio (SNR) for UFSD can be computed as:

$$SNR = \frac{IG}{\sqrt{2q(I_{surface} + I_{bulk}G^2F)\Delta f}} \propto \frac{1}{\sqrt{F}}.\qquad(2.11)$$

Equation (2.11) shows that the introduction of internal gain leads to a reduction of SNR proportional to $\frac{1}{\sqrt{F}}$: if the sensor were the only source of noise, adding internal multiplication would degrade the performances of the whole detector.

In a real detector, however, the situation is different: the read-out electronics contribution dominates the noise. The presence of gain increases the SNR up to the point when the shot noise becomes the dominant noise source. This is shown in Fig. 2.8 (right): as a function of increasing gain, the signal rises linearly while the noise starts increasing only when the shot noise is comparable to the electronic noise.

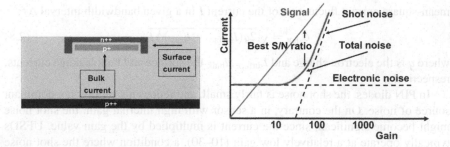

Figure 2.8 *Left*: schematic representation of a UFSD, with the localization of the bulk and of the surface leakage currents. *Right*: signal and shot noise growth as a function of the sensor internal gain.

2.5 TIME-TAGGING DETECTOR

The accurate determination of the time of a hit can happen only if the sensor and the associated front-end electronics match each other's characteristics. On the one hand, the sensor has to provide current signals with a shape that scales, without being distorted, with the amount of energy released by the impinging particle. On the other hand, the electronics has to uniquely identify a given point of the signal, such as a pre-defined voltage value, with the minimum uncertainty.

Figure 2.9 shows a simplified model of a sensor and the associated electronics needed to measure the time of arrival of a particle. For an up-to-date review of current trends in electronics see [114].

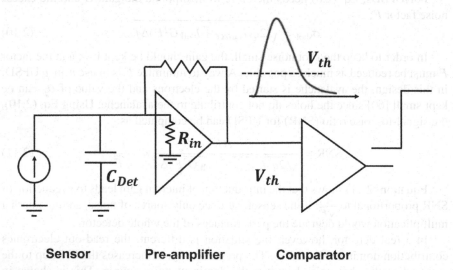

Figure 2.9 Schematic block diagram of a time-tagging detector. The arrival time of a particle is measured when the signal crosses the threshold V_{th} of the comparator.

The sensor, modelled as a capacitance (C_{det}) with a current generator (I_{in}) in parallel, is read out by a preamplifier that shapes the signal. A comparator fires when the preamplifier output exceeds a given voltage value (V_{th}). The output of the comparator is digitized by a Time-to-Digital Converter (TDC). Any effect that changes the shape of the signal near the V_{th} value can anticipate or delay the firing of the comparator, smearing the determination of the time of the hit. The most important factors contributing to the temporal resolution σ_t are shown in Eq. (2.12).

$$\sigma_t^2 = \sigma_{Jitter}^2 + \sigma_{Ionization}^2 + \sigma_{Distortion}^2 + \sigma_{TDC}^2. \tag{2.12}$$

Each of these four terms influences when the signal crosses the discriminator threshold V_{th}. The underlying reason for each of them is:

1. σ_{Jitter}: electronic noise;
2. $\sigma_{Ionization}$: irregularity in the signal shape due to non-uniform energy deposition by the impinging particle. This effect is called Landau noise;
3. $\sigma_{Distortion}$: signal distortion due to non-saturated drift velocity of charge carriers and non-uniform weighting field;
4. σ_{TDC}: the uncertainty due to the finite size of the TDC bin.

These contributions are discussed in detail in the next paragraphs.

2.5.1 JITTER

The presence of noise on the signal, either coming from the sensor, added by the preamplifier electronics or on the discriminator threshold V_{th}, shifts the firing time of the comparator to an earlier or later time.

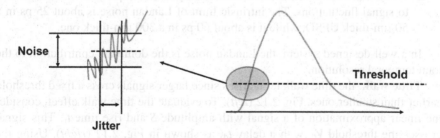

Figure 2.10 The noise causes the early or late firing of the comparator. The uncertainty in time tagging introduced by this effect is called jitter.

This effect, shown in Fig. 2.10, is directly proportional to the noise N and inversely proportional to the slope of the signal around V_{th}. Assuming a constant slope, the slew rate dV/dt can be approximated by the ratio of signal amplitude S over the rise time t_r. This is shown in:

$$\sigma_{Jitter} = \frac{N}{dV/dt} \approx \frac{t_r}{S/N}. \tag{2.13}$$

The minimization of the jitter needs to find the right balance between two opposing aspects. On the one hand, the noise needs to be kept as small as possible while, on the other hand, the slew rate needs to be kept as high as possible. Low noise calls for a small electronics bandwidth while high slew rate requires wide bandwidth.

2.5.2 IONIZATION

A charged particle crossing a silicon sensor generates along its path *e-h* pairs. The *e-h* pairs density varies on an event-by-event basis, producing two effects: (i) non-uniform current signals (Landau noise), and (ii) changes in signal amplitude (time walk). These two effects are related to each other (a discussion of the interplay of the two terms is shown in Section 5.3) since signals with large amplitudes are generated by large localized clusters of charge and have, therefore, also very non-uniform charge depositions. Figure 2.11 shows the simulated signals for a 50 μm-thick UFSD with a gain of about 20. The effects of Landau noise and amplitude variations are visible: the rising edge shows shape variation (Landau noise), and the overall amplitude varies significantly.

Landau noise: non-uniform ionization creates irregularities in the signal shape that represent the physical limit of how uniform the signals from a sensor can be.

The variations of energy deposition are rather large, the signals are not very uniform, and this variability degrades the achievable temporal resolution. There are two methods to mitigate the Landau noise:

1. Integrating the output current over a period of time longer than the typical spike length. This method relies on integration to smooth the signal voltage ramp.
2. Using thin sensors. In a thin sensor, the signal is steeper and is less sensitive to signal fluctuations. The intrinsic limit of Landau noise is about 25 ps in a 50 μm-thick UFSD, while it is about 60 ps in a 300 μm-thick one.

In a well-designed system, the Landau noise is the dominant contribution to the total temporal resolution.

Time walk: the time walk term arises since larger signals cross a fixed threshold earlier than smaller ones, Fig. 2.12 (*left*). To evaluate the time walk effect, consider the linear approximation of a signal with amplitude S and rise time t_r. This signal crosses the threshold V_{th} with a delay t_d, as shown in Fig. 2.12 (*right*). Using the geometrical relationship $t_d/t_r = V_{th}/S$, the time when the signal crosses the threshold can be written as $t_d = t_r V_{th}/S$. The time walk contribution to the temporal resolution is defined as the RMS of t_d:

$$\sigma_{\text{Time walk}} = [t_d]_{\text{RMS}} = [\frac{V_{th}}{S/t_r}]_{\text{RMS}} \propto [\frac{N}{dV/dt}]_{\text{RMS}}. \qquad (2.14)$$

To derive this equation, the relationship $S/t_r = dV/dt$ was used, together with the custom to express V_{th} as a multiple of the noise N of the system. The time walk effect cannot be avoided in systems using a fix discriminator threshold; however, the time walk contribution can be corrected almost completely using appropriate electronic circuits, as explained in Section 2.6.

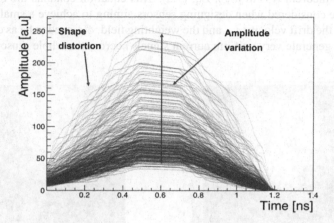

Figure 2.11 WF2 simulation of the signal produced by a MIP in a 50 μm-thick UFSD sensor with an internal gain of about 20.

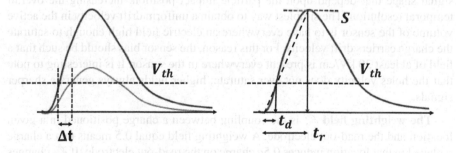

Figure 2.12 *Left*: two concurrent signals with different amplitudes which cross a fixed threshold with a time difference Δt. *Right*: representation of the linear approximation (dashed) used to estimate the V_{th} threshold crossing time t_d, for a signal of amplitude S and rise time t_r.

2.5.3   SIGNAL DISTORTION

In a silicon sensor, the shape of the current signal can be calculated using the Ramo-Shockley's theorem ($i \propto q v_d \mathscr{E}_w$), Eq. (1.5). This equation contains the critical elements to be considered when designing sensors aiming to achieve a small temporal resolution: the drift velocity, v_d, and the weighting field, $\mathscr{E}_w$, need to be as constant as possible to generate very uniform current signals i across the whole sensor surface.

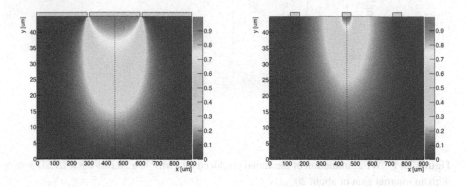

Figure 2.13   The weighting field $\mathscr{E}_w$ maps for a 300 µm pitch sensor with an electrode/gain width of 290 µm (*left*), and 50 µm (*right*). In the narrow implant case (50 µm), the weighting field is not uniform along the x-axis. In this condition, the current $i(t)$ depends on the impact point and its variability increases the temporal uncertainty.

The drift velocity v_d of the charge carriers must be uniform everywhere in the active volume of the sensor. A non-uniform drift velocity induces variations in the signal shape that depend upon the particle impact position, increasing the overall temporal resolution. The simplest way to obtain a uniform drift velocity in the active volume of the sensor is to have everywhere an electric field high enough to saturate the charge carriers drift velocity. For this reason, the sensor bias should be such that a field of at least 30 kV/cm is present everywhere in the sensor. It is interesting to note that the holes velocity does not ever saturate: higher fields always generate sharper signals.

The weighting field $\mathscr{E}_w$ is the coupling between a charge positioned in a given location and the read-out electrode. A weighting field equal 0.5 means that a charge q placed in that location induces $0.5q$ charge on the read-out electrode. If $\mathscr{E}_w$ changes along the pitch implant (the x-axis in the figure), as it is the case for the narrow strip on the *right* side of Fig. 2.13, then the signal shape depends on the hit position of the particle. Implants as large as the pitch, on the other hand, assure the most uniform $\mathscr{E}_w$.

Metalized n^{++} electrode   In a UFSD, the signal, formed in the n^{++} electrode, propagates from the impact point to the read-out electronics. If the propagation of the signal happens in the n^{++} electrode, *left* side of Fig. 2.14, a non-zero delay is accumulated due to the n^{++} resistivity. This delay, albeit rather small (~ 0.5 ps/μm), can spoil the temporal resolution. For this reason, UFSDs are often fully metallized since propagation in the metal does not yield to a significant delay. In most UFSD design, the metal is not directly in contact with the n^{++} electrode but is placed over and oxide, *right* side of Fig. 2.14. The UFSD readout, therefore, is a mixed AC- and DC-coupled system: the signal is AC-coupled to the metal to avoid propagation delay, and the n^{++} electrode and the metal are in contact at the edge of the pad to avoid a bipolar signal, typical of an AC-coupled readout scheme.

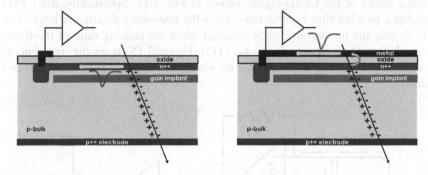

Figure 2.14 Signal propagation on the UFSD surface for a pad without (*left*), and with (*right*) metal.

The preceding three points indicate a simple receipt: the optimum sensor geometry resembles as much as possible that of a parallel plate capacitor, with the pitch much larger than the sensor thickness. Under these conditions, the electric and weighting fields are as uniform as they can be, assuring, as a function of the hit position, constant drift velocity and coupling between a charge and the read-out electrode. The metalized surface avoids position-dependent delay.

2.5.4 TDC

The TDC records the time of the discriminator firing in a time bin of finite width ΔT, given by the TDC least significant bit. Therefore, this process adds a contribution to time uncertainty equal to $\Delta T / \sqrt{12}$. Thanks to the fine binning of the TDCs commonly used in high-energy physics experiments, for instance, the High Precision TDC [104] with a bin width of 25 ps, the σ_{TDC} term can be neglected.

2.6 UFSD READ-OUT ELECTRONIC

The temporal performances of a detector using UFSDs depend on the accurate matching of the sensor to the read-out electronics. The sensor should provide large signals, with a shape as constant as possible, while the electronics should minimize the jitter contribution. A key constraint in the design of the electronics is the amount of power available. Past read-out chips designed for timing applications, for example those reported in [25, 27], proved that a temporal resolution of 50 ps could be reached using ~ 30 mW per channel. In the past few years, driven by the use of UFSDs in the ATLAS and CMS experiments at CERN, two dedicated ASICs, the ALTIROC [22] and the ETROC [73], have been designed to reach a jitter term lower than 30 ps for pads of capacitance 2–6 pF while using only 2–3 mW/channel. Independently, the FAST ASIC family [24] has been developed to study low-power (~ 1.3 mW) front-end architectures suitable to reach also about 30 ps temporal resolution when coupled to a 50 μm-thick UFSD with 2–6 pF of capacitance.

The design of an ASIC tailored to read UFSDs sensors needs to consider the particular shape of the UFSD signal, shown in Fig. 2.12. Specifically, the UFSD signal has a peaking time $t_{p,s}$ that determines the interesting bandwidth interval of the front-end: the minimum jitter is obtained when the peaking time of the front-end equals that of the sensor, $t_{p,fe} = t_{p,s}$ [115]. Figure 2.15 shows the most important quantities that need to be considered when matching a UFSD to a front-end amplifier.

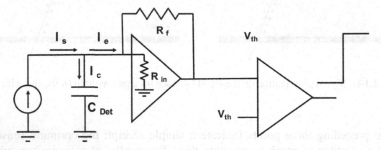

Figure 2.15 A sketch showing the key parameters to be considered in the matching of a silicon sensor to its front-end. The silicon sensor is represented as a current source with a capacitor in parallel, while the front-end is characterized by an input impedance R_{in} and a feedback resistance R_f.

The key point in the design of the first stage of the ASIC is to decide what part of the current signal I_s flows into the front-end, I_e, and what part is integrated on the sensor capacitance, I_c. This split is controlled by two time constants: the signal duration τ_s and the system discharge time $\tau_{in} = C_{det}R_{in}$. If $\tau_{in} < \tau_s$ then $I_e \sim -I_s$ and $I_c \sim 0$. In this configuration, the amplifier works in current mode (current-mode amplifiers), and the amplifier output has the same shape as the input current signal. Considering that the typical UFSD value of τ_s is ~ 1 ns for a 50 μm-thick UFSD and a capacitance of $C_{det} = 4$ pF, the front-end should be designed with R_{in} lower than 250 Ω. On the other hand, if the capacitance of the sensor is large or the input impedance cannot be small, the current is integrated into the sensor capacitance. In this condition, the front-end sees a voltage signal (voltage-mode amplifiers) whose maximum amplitude depends on the discharge time τ_{in}. Assuming that the signal is completely integrated over C_{det}, a 50 μm-thick UFSD with a gain of 10 and $C_{det} = 4$ pF generates a voltage signal of about 1.25 mV. Generally, current-mode amplifiers are the best choice for preserving both the rising and falling edges of the fast sensor pulse generated by a UFSD. However, this is feasible only with small values of C_{det}. The optimum bandwidth for a current-mode amplifier reading a 50 μm-thick UFSD is in the range of 400–800 MHz.

The amplitude of the output signal is controlled by the feedback resistance R_f. Its value cannot be made too large since R_{in} increases with R_f, which, if too large, degrades the sensor-front-end matching. For amplifiers with an open-loop gain of 40 dB, a good trade-off is achieved using $R_f \sim$ 5–20 kΩ. In this case, R_{in} can be reduced by maximizing the input transistor transconductance g_m.

2.6.1 TIME WALK CORRECTION

In a silicon detector, the distribution of the signal amplitude generated by a MIP follows a Landau probability function: most of the signals are clustered around the most probable value, while there is also the possibility to have much larger signals. If the front-end electronics uses a comparator with a fixed threshold, see Fig. 2.11, large signals appear to arrive earlier than small signals. This effect, called time walk, spoils the temporal resolution of the detector completely. For this reason, it is necessary to introduce a time walk correction mechanism.

There are two common approaches used to assign the time, see Fig. 2.16, and eliminate the time walk effect: (i) the Constant Fraction Discriminator (CFD) and (ii) the Time-over-Threshold (ToT).

1. The CFD method sets the time of arrival of the particle not when the signal amplitude crosses a fixed threshold but at a fixed fraction of the amplitude. This method works well under the assumption that signals of different amplitudes scale without any distortion.

2. The ToT method uses two points in time to determine the time of arrival of a particle. The time t_1, when the signal exceeds the threshold, is corrected with a function of the time-over-threshold quantity $t_2 - t_1$, which depends on the signal amplitude.

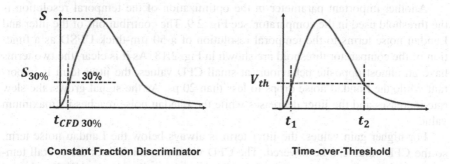

Constant Fraction Discriminator　　　**Time-over-Threshold**

Figure 2.16 Time-tagging techniques. *Left*: Constant Fraction Discrimination, the time assigned scales in first approximation with the amplitudes of the signals. *Right*: Time-over-Threshold, the time t_1 is time walk corrected using a function of the quantity $t_2 - t_1$, which is proportional to the signal amplitude.

2.7 UFSD TEMPORAL RESOLUTION

The correct matching of a UFSD with a front-end circuit able to exploit its signal characteristics leads to excellent temporal resolutions. Figure 2.17 shows the WF2 simulation of the temporal resolution as a function of sensor thickness for a system composed of a UFSD matched to an amplifier working in current mode. The gain of the sensor is assumed to be about 20 and its capacitance to be 6 pF. The plot shows the jitter and Landau noise contributions and their sum in quadrature. Both factors decrease with the sensor thickness, and they become similar for thicknesses below 100 μm. A 50 μm-thick UFSD achieves a temporal resolution of 30–40 ps. This simulation highlights an essential feature of the UFSD design: the Landau noise limits the achievable temporal resolution, and this limit is a function of the sensor thickness.

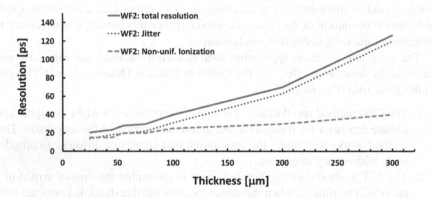

Figure 2.17 WF2 simulation (gain = 20, C_{det} = 6 pF) of the jitter and Landau noise terms as a function of the UFSD thickness.

Another important parameter in the optimization of the temporal resolution is the threshold used in the comparator, see Fig. 2.9. The contributions of the jitter and Landau noise terms to the temporal resolution of a 50 μm-thick UFSD as a function of the comparator threshold are shown in Fig. 2.18. As it is clear, the two terms have an almost opposite behaviour: at small CFD values, the jitter term is important while the Landau noise drops to less than 20 ps. As the signal grows, the slew rate increases and the jitter decreases while the Landau noise reaches its maximum value.

For higher gain values, the jitter term is always below the Landau noise term, so the CFD value can be lowered. The CFD value that minimizes the overall temporal resolution is normally in the 0.1–0.3 range, depending on the UFSD gain and electronic noise.

The key ingredient to achieving an excellent temporal resolution is the gain mechanism. However, a given value of gain, for example, gain = 15, results in very different temporal resolutions according to what bias voltage is used to achieve it.

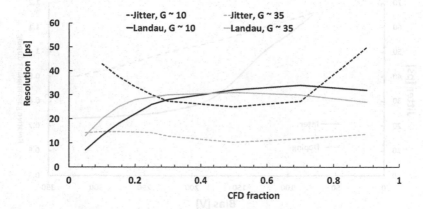

Figure 2.18 WF2 simulation of the jitter and Landau noise terms as a function of the CFD value, for two different internal gains of a 50 μm-thick UFSD.

Figure 2.19 shows the simulated jitter of a 50 μm-thick UFSD with gain 15, as a function of different combinations of external bias and gain implant doping.

The minimum jitter is obtained using the highest bias voltage. In this condition, the electrons drift velocity is saturated, and the holes drift velocity is at the highest value. Both factors contribute in sharpening the signal. If the gain implant doping increases, the bias voltage has to drop to keep the field in the gain layer constant. At high relative gain implant doping, the gain of 15 is reached at a relatively low bias voltage, and the temporal resolution worsens significantly. The correct doping range for the gain implant is where the jitter term remains almost constant, in this example between the values 0.98 and 1.1.

Figure 2.20 shows the evolution of the total current induced in 50 μm-thick UFSDs with gain = 10, as a function of bias voltage, simulated with WF2. The plot sequence shows the importance of the bias setting: the higher the bias, the sharper and shorter the signal becomes and, consequently, the smaller the temporal uncertainty.

2.8 BUILDING BLOCKS OF MULTI-PADS UFSD

The final goal in the design of UFSD devices is to produce large-area sensors with hundreds of pixels. This development has been accelerated by the decisions of the two CERN experiments ATLAS [8] and CMS [9] to instrument a timing layer using UFSDs.

The geometry chosen by the ATLAS and CMS experiments are very similar: ATLAS (CMS) will use sensors with 15×30 (16×32) pads, each pad being 1.3×1.3 mm^2. The design of a large, multi-pad UFSD presents many challenges in terms of uniformity, stability, and the design of the region between pads. The area between pads is particularly delicate, as it might lead to noise and premature breakdown.

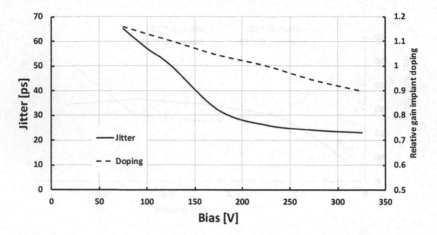

Figure 2.19 WF2 simulation of the jitter (left *y*-axis) of a 50 μm-thick UFSD for different combinations of gain implant doping (right *y*-axis) and external bias voltage to achieve a gain of 15.

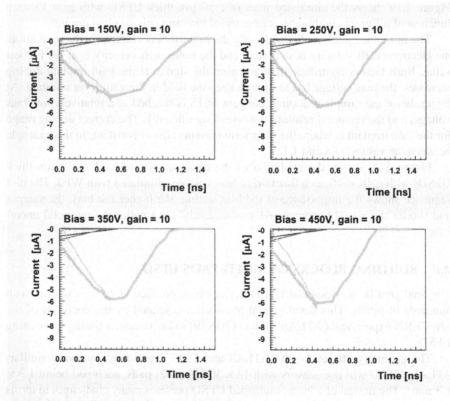

Figure 2.20 Total current (light grey bold in the plots), as predicted by WF2, in 50 μm-thick UFSDs with fixed gain = 10, as a function of the bias voltage.

Figure 2.21 shows a simplified cross-section of a portion of a multi-pad UFSD (not-to-scale). At the physical edge of the sensor, there are the guard-rings. These structures have the task of grading the voltage from the sensor edge, held at the bias voltage, to the first read-out pad, held at virtual ground by the read-out electronics.

Each guard-ring consists of an n^{++} doped implant, equipped with metal field plates. A p^+ implant (either a p-stop or p-spray) is (in most designs) interposed between each pair of guard-rings, with the outer ones left floating, and the inner one generally grounded in order to collect the leakage current generated outside the core region of the device. The task of the guard-rings becomes progressively more difficult as the sensor thickness decreases. In a silicon sensor, the depletion region extends laterally by a distance similar to the sensor thickness. For this reason, the floating guard-rings need to be placed within a distance of about two to three sensor thickness from the inner one. As an example, in a 50 μm-thick UFSD, the guard-rings should extend laterally by no more than 100–150 μm. Decreasing the lateral spread of the guard-rings increases the electric field between them, leading to possible breakdowns. This is a challenging problem in the quest to design very thin (~ 20–30 μm) UFSDs. It is also possible to leave floating even the inner guard-ring: in this configuration, the pads next to the guard-ring will have a higher current.

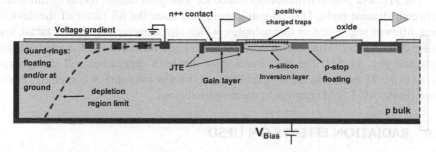

Figure 2.21 Cross cut of a multi-pads UFSD (not to scale) with a schematic view of the building blocks of the device. From the device physical edge: guard-rings, pad with JTEs, inter-pad region with p-stop.

The gain region, as shown on the *left* side of Fig. 2.1, is surrounded by a deep n^{++} implant called Junction Termination Extension (JTE), equipped with metal field plate. The JTE is located around each pad and it ensures that the e-h pairs generated by particles impinging in the region between the pads do not reach the gain layer. When the impact point is where the gain layer is implanted, the electrons initiate the multiplication mechanism without delay. On the contrary, if the e-h pairs are generated in the inter-gap region, the electrons will have to drift to the gain implant, and the multiplication process will start with a considerable delay, see Figure 2.22, left side. Considering a drift velocity of 10 ps/μm, the multiplication can easily be delayed by hundreds of picoseconds, causing a completely wrong assignment of the particles time of arrival. This problem can be avoided by inserting the JTE at the periphery of each pad, as shown on the *right* part of the picture. Therefore, the JTE structure delimits the active area of the sensor to the regions where the gain implant is present.

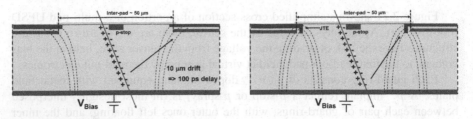

Figure 2.22 Sketch of the inter-pad region. The dashed lines show the *e-h* pairs drift lines. In a design without JTE (*left*), the *e-h* pairs generated in the inter-pad will reach the gain layer after a long drift. On the contrary, when the JTE is present (*right*), these charges are collected without reaching the gain layer.

Common to every silicon sensor using the n-in-p design, the n^{++} implants need to be isolated from each other by an extra p^{++} implant, the so-called p-stop. The p-stop has a doping concentration in the range 10^{15}–10^{17}/cm^3. Without p-stops, the inversion layer created by the positive charges at the Si-SiO$_2$ interface would short together all n^{++} implants.

The JTE and p-stop terminations introduce a no-gain region, tenths of μm wide, between adjacent pads. This no-gain region decreases the fill factor of the device. Two different segmentation technologies, which aim at increasing the fill factor, are under development: Trenches Isolation (TI) [39] and Resistive AC-Coupled Silicon Detectors [61, 54, 53]. The first results show that both approaches will strongly reduce (in the TI approach, the no-gain distance is a few microns) or completely eliminate (in the AC-LGAD approach) the no-gain region.

2.9   RADIATION EFFECTS ON UFSD

In order to extend the range of applications of UFSDs, their radiation resistance has been studied in detail. In the past few years, several design improvements, such as carbon co-implantation and the deep gain implant, have extended the maximum fluence at which UFSDs are used well above $1 \cdot 10^{15}$ n$_{eq}$/cm^2. The two first large-scale applications of UFSDs, the timing layers in the ATLAS and CMS experiments, require that the sensors continue to work up to fluences of about 2–$3 \cdot 10^{15}$ n$_{eq}$/cm^2.

The main effects of radiation damage on PIN silicon sensors have been discussed in Section 1.3. In this paragraph, the effects of radiation damage specific to the UFSDs will be discussed.

2.9.1   INCREASED LEAKAGE CURRENT: POWER CONSUMPTION AND SHOT NOISE

In sensors with internal gain, the leakage current generated in the bulk is multiplied by the gain factor G before being collected at the electrodes. This unavoidable effect increases power consumption:

$$i_{\text{gain}} = G \cdot i_{\text{no-gain}} \tag{2.15}$$

$$P_{\text{gain}} = G \cdot P_{\text{no-gain}}. \tag{2.16}$$

Power consumption can be reduced using thin sensors since both leakage current and operating voltage are lower, and by cooling the sensor, since the leakage current depends on the temperature as:

$$i(T) \propto T^2 \exp \frac{1.2\ eV}{2\ k_B T}, \tag{2.17}$$

where k_B the Boltzmann's constant and T is expressed in Kelvin. A temperature variation of 7 degrees leads to a current variation of a factor of two; cooling a sensor to $T = -30\ °C$, the operating temperature in ATLAS and CMS, decreases the current by about a factor of 100 with respect of room temperature.

A second significant effect due to the high leakage current in irradiated sensors is the shot noise increase. Shot noise, see Section 2.4, is usually lower than the electronic noise in unirradiated sensors, but it might become the dominant source of noise in irradiated ones. As an example of the impact of the shot noise, Fig. 2.23 shows the shot noise as a function of the irradiation fluence for a 1.3×1.3 mm^2 50 µm-thick UFSD, assuming an analog bandwidth of 500 MHz, and $k = 0.2$, where $k = \alpha_p/\alpha_n$ is the ratio between the impact ionization coefficients of holes and electrons. These conditions are similar to those of the ATLAS and CMS timing layers. The *top* plot shows the impact of the temperature on the shot noise. Clearly, strong cooling is necessary to keep the noise low. The *bottom* plot illustrates the evolution of the total noise in the ALTIROC and ETROC ASICs as a function of fluence. Both ASICs have an equivalent noise charge (ENC) of about 1400 electrons: even at a fluence of $5 \cdot 10^{15}$ n$_{eq}$/cm^2, the noise increase due to shot noise is rather mild. In conclusion, Fig. 2.23 suggests that operating irradiated sensors at low gain and low temperature is the key to maintain the shot noise below the electronic noise.

2.9.2 VARIATION IN DOPING CONCENTRATION: LOSS OF GAIN AND HIGHER DEPLETION VOLTAGE

Another important consequence of irradiation is the variation of the doping concentration as a function of the fluence, Eq. (1.9). This mechanism, discussed in Section 1.3, consists of two opposite and concurrent contributions: initial acceptor removal and acceptor creation. Both mechanisms apply to the gain implant and the bulk of a UFSD. Figure 2.24 shows the evolution of the boron concentrations as a function of fluence in the gain implant and bulk, with initial boron densities of $3 \cdot 10^{16}$ atoms/cm^3 and $5 \cdot 10^{12}$ atoms/cm^3, respectively. At sufficiently high fluences ($\Phi > 1 \cdot 10^{16}$ n$_{eq}$/cm^2), the acceptor density of the gain implant matches the bulk acceptor density, indicating a complete removal of the initial gain implant doping.

Using the evolution of the density of acceptors in the gain implant and bulk shown in Fig. 2.24, it is possible to calculate the full depletion voltage of a UFSD

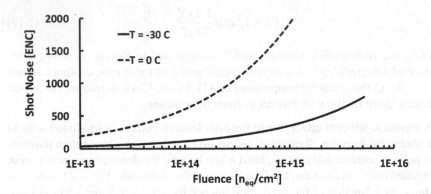

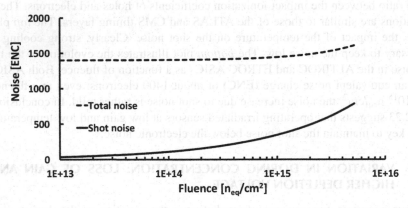

Figure 2.23 Noise evolution as a function of the irradiation fluence, for a 1.3×1.3 mm^2 50 μm-thick UFSD (gain = 20, bandwidth = 500 MHz, $k = 0.2$). *Top*: shot noise as a function of the irradiation fluence for two different values of temperature. *Bottom*: total and shot noise as a function of fluence ($T = -30\,°C$).

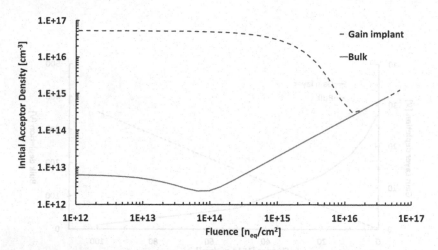

Figure 2.24 Evolution of the boron density for gain implant and p-doped bulk, computed using Eq. (1.9), as a function of the irradiation fluence, with initial boron densities of $3 \cdot 10^{16}$ atoms/cm^3 and $5 \cdot 10^{12}$ atoms/cm^3, respectively.

as a function of the fluence. This value is the sum of two terms: the depletion voltage of the gain layer and the depletion voltage of the bulk. The evolution of the full depletion voltage for a 50 μm-thick UFSD is shown in Fig. 2.25.

As shown in Section 1.3.1, irradiation displaces silicon atoms outside the lattice that deactivates the boron via kick-out reactions, Fig. 2.26 (*top*). This interpretation of the acceptor removal mechanism motivated two research lines within the RD50 collaboration [113], with the goal of mitigating the deactivation of the gain implant.

The first research path aims at reducing the concentration of interstitial silicon defects. This is achieved by implanting carbon atoms in the volume of the gain implant. Carbon replaces boron in the ion-defect complexes formation, Fig. 2.26 (*middle*), so the boron deactivation is mitigated. Also, carbon atoms in substitutional position tend to pair with boron interstitials and form centres with energy approximately 80% of the boron acceptor level energy [4]. The second research path consists of replacing the boron atoms with gallium, which, being heavier, are predicted to have a slower acceptor removal, Fig. 2.26 (*bottom*).

2.9.3 CHARGE TRAPPING ON THE UFSD SIGNAL SHAPE

In irradiated sensors, the charge carriers are subjected to trapping, Section 1.3, a mechanism that leads to charge collection decrease and affects the output signal shape. Trapping increases with the irradiation fluence and with the drift time of the charge carriers. For irradiation fluence of the order of 10^{15} n$_{eq}$/cm^2, the trapping time is ~ 2 ns, roughly three times the drift time of the charge carriers in a 50 μm-thick UFSD.

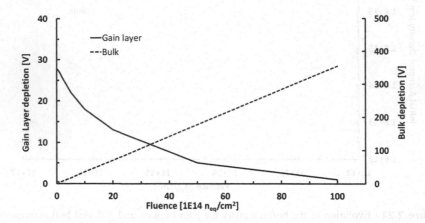

Figure 2.25 Evolution of the depletion voltage of the gain layer and bulk as a function of fluence, for a 50 μm-thick UFSD.

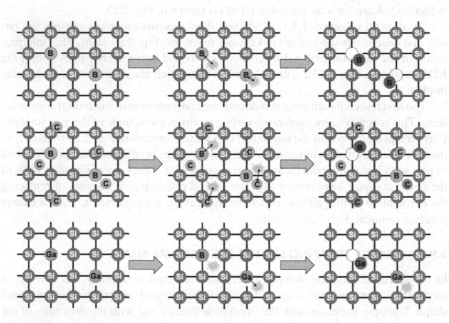

Figure 2.26 Sketch of the acceptor removal mechanism for gain implants with different dopants: boron (*top*), boron enriched with carbon (*middle*), and gallium (*bottom*).

The effects of charge trapping on the current signal of a 50 μm-thick UFSD at two irradiation levels, $\Phi = 1 \cdot 10^{15}$ n_{eq}/cm^2 and $\Phi = 2.5 \cdot 10^{15}$ n_{eq}/cm^2, are presented in Fig. 2.27. Here, the hypothetical signals in absence of trapping are compared with those when trapping is present.

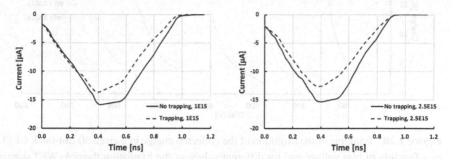

Figure 2.27 WF2 simulation of the signal shape from a 50 μm-thick UFSD, with and without trapping. *Left:* Current signals with and without trapping at $\Phi = 1 \cdot 10^{15}$ n_{eq}/cm^2. *Right:* Current signals with and without trapping at $\Phi = 2.5 \cdot 10^{15}$ n_{eq}/cm^2.

Even at $\Phi = 2.5 \cdot 10^{15}$ n_{eq}/cm^2, the signal does not change dramatically; it is smaller but still usable. Interestingly, the rising edge of the signal, the key feature used for timing measurement, is not affected much by trapping.

2.10 GAIN RECOVERY IN IRRADIATED UFSD

Due to the acceptor removal effect, the gain in UFSD decreases rapidly with increasing fluence. Consider a 50 μm-thick UFSD with a gain of about 20 at a bias voltage of $V_{bias} = 110$ V. For unchanged bias voltage, after a fluence of $\Phi = 4 \cdot 10^{14}$ n_{eq}/cm^2, the gain is reduced to about 7, while at $\Phi = 8 \cdot 10^{14}$ n_{eq}/cm^2, it is reduced to less than 4. The decrease of the gain is due to the reduction of the electric field in the gain layer caused by the gain implant's inactivation. This field loss can be compensated by raising the bias voltage, as shown in the WF2 simulation of Fig. 2.28. In this example, before irradiation, the gain layer generates about 90% of the electric field. After a fluence of $\Phi = 8 \cdot 10^{14}$ n_{eq}/cm^2, this fraction is reduced to about 60%, and at $\Phi = 1.5 \cdot 10^{15}$ n_{eq}/cm^2 becomes 45%. When the bias voltage reaches about 500–750 V over 50 μm ($\mathscr{E} \sim 100$–150 kV/cm), the multiplication mechanism also begins in the bulk. This effect is called *bulk gain*. Bulk gain has a very sharp turn-on, and it leads rapidly to a breakdown condition.

Another parameter that affects the gain recovery mechanism is the depth of the multiplication implant. As discussed in Section 2.1, the same gain in UFSDs with shallow or deep gain implants is achieved at different electric fields and, consequently, with different mean free paths λ, Eq. (2.2). Figure 2.29 *(top)* shows the mean free path λ as a function of the electric field $\mathscr{E}$. The value of the electric field in a shallow (deep) gain implant is ~ 300 kV/cm (~ 400 kV/cm). As the field generated by the gain layer drops due to irradiation, λ increases. When the bias voltage is

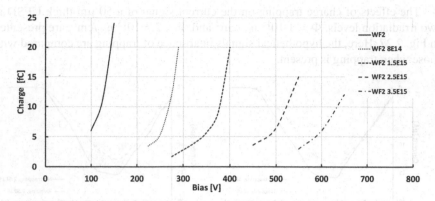

Figure 2.28 Example of the evolution of the collected charge in a given 50 μm-thick UFSD, as a function of bias voltage and for different values of the irradiation fluence (WF2 simulation).

increased, the value of λ is shortened, and the gain is restored. However, the effect on λ of a given bias increase is not constant, it depends on the value of the electric field. This effect is shown in the *bottom* part of Fig. 2.29 plotting $d\lambda/d\mathscr{E}$ as a function of $\mathscr{E}$. Since $d\lambda/d\mathscr{E}$ is larger for a deep gain implant (low field working point), this design has a higher gain recovery capability than the design that employs a shallow gain implant (high field working point).

2.11 ADDITIONAL LGAD DESIGNS

In this short section, two additional LGAD designs are briefly presented. These designs are included only for completeness, as, up to now, their capabilities of tagging time accurately (30 to 50 ps) have not been demonstrated yet.

2.11.1 DOUBLE-SIDED LGAD DESIGN

The segmentation technique presented in Section 2.8 is clearly non-ideal, as it requires the introduction of several structures (the JTE and p-stops) and leads to a no-gain area between pads. A possible alternative design uses a double-sided production process, where the cathode side has an un-interrupted gain layer, and the pixellation is obtained on the ohmic side of the junction [31]. Figure 2.30 shows on the *left* side the standard n-in-p UFSD design, while on the *right* side, the double-sided one. In both sketches, the read-out electronics is at the ground potential, while the bias voltage is negative on the *left* sketch and it is positive on the *right* side. The design of the gain layer is identical in both designs; an acceptor p^+ implant in a p-doped bulk, the avalanche is started by the drifting electrons, and, if the electric field is well tuned, the holes do not contribute to the multiplication process. The FBK UFSD1 production, see Appendix A, is double-sided, and it uses 300 μm-thick active sensors.

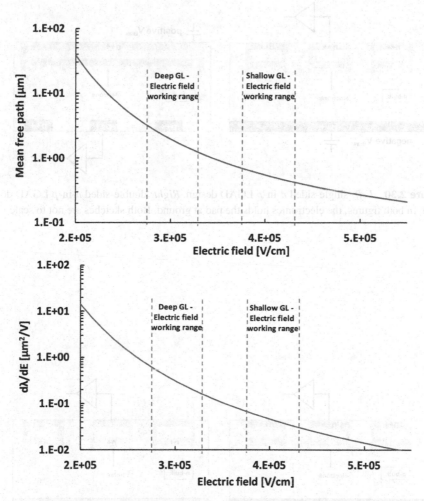

Figure 2.29 Electrons mean free path between two subsequent scattering events producing secondary charges λ (*top*) and $d\lambda/d\mathcal{E}$ (*bottom*) at $T = 300$ K as a function of the electric field $\mathcal{E}$, according to the Massey impact ionization model.

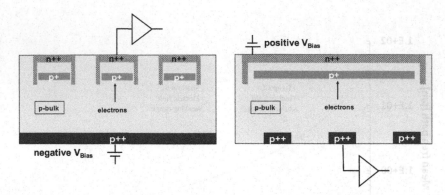

Figure 2.30   *Left:* single-sided *n*-in-*p* LGAD design. *Right:* double-sided *n*-in-*p* LGAD design. In both figures, the electronics holds the pad at ground. Both sketches are not to scale.

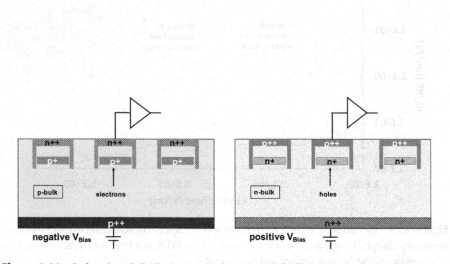

Figure 2.31   *Left:* *n*-in-*p* LGAD design. *Right:* *p*-in-*n* LGAD design. Both sketches are not to scale.

The main drawback of the double-sided design is the impossibility of using a thin active volume. The need to process the wafer on both sides forces the use of thick sensors (200–300 μm) that can be flipped during manufacturing. For this reason, the temporal resolution of the double-sided design is rather poor: about 100 ps.

2.11.2   n-in-p VS p-in-n LGAD DESIGN

The properties of the LGAD design presented so far are based on the n-in-p sensor structure: the bulk and the gain implants are p-doped, and the multiplication mechanism is started by the electrons entering the gain layer. This design is shown on the *left* side of Fig. 2.31. It is possible to design an LGAD using the p-in-n architecture, as shown on the *right* side of Fig. 2.31. In this configuration, the bulk and the gain implants are n-doped, and the holes start the multiplication process. The p-in-n design is intrinsically less stable than the n-in-p design: the electric field necessary to start impact ionization with electrons is lower than with holes (due to the smaller impact ionization coefficient, see Section 2.1.1), so in the n-in-p design only the electrons generate multiplications, while in the p-in-n design both electrons and holes do. The p-in-n design, therefore, can easily lead to uncontrolled avalanche multiplication. An additional drawback of the p-in-n design is the longer signal rise time, controlled by the holes drift velocity.

3 Numerical Modelling and Simulation

3.1 INTRODUCTION

Modern silicon technologies, based on highly complex systems, require rigorous methods to predict the behaviour of electronic devices and the computational approach becomes essential. In the past decades, the applied mathematics branch called numerical computation has closely followed the evolution of information technologies, refining its techniques according to the complexity of the systems to be described and benefiting from the upgrade in the computational hardware performances. In this view, one of the most significant achievements is represented by the introduction of the Finite Element (FE) analysis, based on a particular solving strategy for partial differential equations (PDE), which has been developed from the 30s of the last century.

FE analysis is a mathematical tool able to offer a considerable saving of computational costs and the advantage of setting a proper approximation level with respect to the exact solution. The user can choose the optimal balance between the precision of the result and the time required to process data.

The important ingredients of any physical modelling of semiconductor devices are presented in the following sections. Before focusing our attention on the Drift-Diffusion (DD) framework, one of the most commonly used techniques both in the industry and in the R&D field, the microscopic description of charge carriers transport, as well as its numerical treatment, will be provided. To this aim, how these models are discretized before being implemented in a calculator will be briefly described. Finally, the case study of a UFSD-based detector is presented at the end of the chapter to show how theory applies to real life. In this section, standard carrier statistics and transport is coupled with advanced semiconductor physics (quantum models and radiation effects) to highlight the properties and issues of simulating the electrical behaviour and the operating performances of a real device.

3.2 PHYSICAL MODELLING OF SEMICONDUCTOR DEVICES

As anticipated in the introduction, the ingredients of physics-based modelling are now introduced. The first ingredient is the relationship between the electric field and the charge density, described by the Poisson's equation. The second ingredient is how the charge carriers, i.e., electrons and holes, *react* to the applied field or, in other words, the carrier dynamics. This step needs a dedicated transport model (TM). Here, two TMs models are detailed: the Boltzmann Transport Equation (BTE) and the Drift-Diffusion (DD).

3.2.1 ELECTROMAGNETIC MODEL: THE POISSON'S EQUATION

The *Poisson's equation* connects the electric field acting within a semiconductor material $\mathscr{E}$ and the overall charge density ρ (the sum of the densities of positive, negative and fixed charges):

$$\frac{\mathrm{d}^2 U_x}{\mathrm{d}x^2} = q\frac{\rho}{\varepsilon}, \tag{3.1}$$

where U_x is the potential energy along an hypothetical direction x and ε is the dielectric constant of the semiconductor. To make Eq. (3.1) explicit with respect to the electric field, it is sufficient to remember that the potential energy and the electrostatic potential φ are linked by

$$U_x = -q\varphi_x \tag{3.2}$$

and that

$$\mathscr{E}_x = -\frac{\mathrm{d}\varphi_x}{\mathrm{d}x}, \tag{3.3}$$

so it can be written

$$\frac{\mathrm{d}\mathscr{E}_x}{\mathrm{d}x} = \frac{\rho}{\varepsilon}, \tag{3.4}$$

which is the most common way to represent the Poisson's equation for semiconductors.

Depending on the system to be modelled, Eq. (3.4) has to be rewritten as a function of two or three spatial variables, transforming the derivative into a divergence and the density ρ into a distribution of charges $\rho(\mathbf{r})$ in the generic space vector $\mathbf{r}$. In the following section of this chapter, it will be shown how to write Eq. (3.1) in a more general form and, especially, how to solve it within a FE framework.

3.2.2 TRANSPORT MODELS

The simplest Transport Model suitable for electronic devices modelling in the semi-classical approach is the so-called *Boltzmann Transport Equation* (BTE). Written for a generic distribution function $f(\mathbf{k}, \mathbf{r}, t)$, which describes a population of charge carriers, the BTE is

$$\frac{\partial f}{\partial t} + \mathbf{v}(\mathbf{k}) \cdot \nabla_{\mathbf{r}} f + \frac{\mathbf{F}}{\hbar} \cdot \nabla_{\mathbf{k}} f = \left.\frac{\mathrm{d}f}{\mathrm{d}t}\right|_{\text{coll}}, \tag{3.5}$$

where $\mathbf{v}(\mathbf{k})$ is the fermion group velocity. In the equation above, the term

$$\mathbf{F}\left(n, p, \varphi, \nabla, \frac{\partial}{\partial t}\right) \tag{3.6}$$

is a hypothetical force acting on the system, written as a function of the electron-hole carrier densities n and p, the electrostatic potential φ, and the gradient and derivative operators.

The last term of the BTE is called *collision term*, and describes the dynamics of the perturbation induced by the force **F**. Under the *relaxation time* approximation, it can be written

$$\left.\frac{df}{dt}\right|_{coll} = \frac{|f(\mathbf{k},\mathbf{r},t) - f_0(\mathbf{k})|}{\tau(\mathbf{k})}, \tag{3.7}$$

where $f_0(\mathbf{k})$ is the distribution function at equilibrium and $\tau(\mathbf{k})$ the time required to restore such equilibrium (relax) after the initial perturbation. Following the statistical theory,

$$f(\mathbf{k},\mathbf{r},t)\,\Delta\mathbf{k}\,\Delta\mathbf{r}\,\Delta t \tag{3.8}$$

represents the number of electrons/holes having momentum $\mathbf{p} = \hbar\mathbf{k}$, at position $\mathbf{r}$ and time t. Similarly, integrating Eq. (3.8) with respect to $\mathbf{r}$, the number of electrons/holes having momentum $\mathbf{p} = \hbar\mathbf{k}$ at time t is found and, finally, by integrating with respect to $\mathbf{k}$, the number of carriers at position $\mathbf{r}$ and time t is determined. Introducing now an opportune function $\lambda(\mathbf{k})$ such that

$$\lambda(\mathbf{k}) = \sum_{j=0}^{N} a_j \mathbf{k}^j, \tag{3.9}$$

with

$$a_0 = 0, \quad a_1 = \hbar\mathbf{k}, \quad a_2 = \frac{\hbar^2\mathbf{k}^2}{2m^*}, \quad \ldots \tag{3.10}$$

then a set of N *moments* M_j of the distribution f having the general form

$$\mathsf{M}_j = \int \lambda_j(\mathbf{k})\, f(\mathbf{k},\mathbf{r},t)\, d\mathbf{k} \tag{3.11}$$

are found. These moments assume a noticeable relevance since they provide information about the properties of the system. For instance, applying the 0th-order moment M_0 to the BTE written for the electrons, the following equation

$$\frac{\partial n(\mathbf{r},t)}{\partial t} + \nabla_{\mathbf{r}}(\langle v_n \rangle\, n(\mathbf{r},t)) = \left.\frac{dn(\mathbf{r},t)}{dt}\right|_{coll} \tag{3.12}$$

is obtained, which represents the *continuity equation* (i.e., at the same time, a charge conservation law and a transport equation for electrons), since

$$\mathsf{M}_0 = \int f(\mathbf{k},\mathbf{r},t)\, d\mathbf{k} = n(\mathbf{r},t). \tag{3.13}$$

Note that, in order to ensure that Eq. (3.12) is valid, it is assumed that

$$\langle v_n \rangle = \frac{\int \mathbf{v}(\mathbf{k})\, f(\mathbf{k},\mathbf{r},t)\, d\mathbf{k}}{\int f(\mathbf{k},\mathbf{r},t)\, d\mathbf{k}} \tag{3.14}$$

is the average electron velocity or, to simplify, the electron *drift velocity* v_n. Since the electron current density can be expressed as

$$J_n = -q\,v_n\, n(\mathbf{r},t), \tag{3.15}$$

and assuming that

$$\left. \frac{dn(\mathbf{r},t)}{dt} \right|_{coll} = -U_n(\mathbf{r},t),$$ (3.16)

where the term U_n will be defined shortly, the BTE for electrons becomes

$$\frac{\partial n(\mathbf{r},t)}{\partial t} = \frac{1}{q}\nabla_\mathbf{r} J_n(\mathbf{r},t) - U_n(\mathbf{r},t)$$ (3.17)

and, similarly for holes

$$\frac{\partial p(\mathbf{r},t)}{\partial t} = -\frac{1}{q}\nabla_\mathbf{r} J_p(\mathbf{r},t) - U_p(\mathbf{r},t).$$ (3.18)

What has been done so far is to start from the generic expression of the BTE and, introducing the method of moments, rewrite it in a form more suitable to describe the transport of free charges in semiconductors. Such a procedure can be improved by adding a couple of other considerations. First of all, recall that in solid-state physics the current density of charge carriers $J_{n,p}$ has a contribution driven by the electric field (called drift current) and a second component due to the gradient of charge density (the diffusion part). So, in one space dimension, it can be written as

$$J_n = q\mu_n n\mathscr{E} + qD_n \frac{\partial n}{\partial x}$$
$$J_p = q\mu_p p\mathscr{E} - qD_p \frac{\partial p}{\partial x}$$ (3.19)

where $\mu_{n,p} = v_{n,p}/\mathscr{E}$ are the electron-hole mobilities and

$$D_{n,p} = \mu_{n,p} k_B T$$ (3.20)

the Einstein diffusion coefficients, functions of the material-dependent mobilities $\mu_{n,p}$, the Boltzmann constant k_B and the absolute temperature T. Finally, the term $U_{n,p}$ is the so-called net generation-recombination (GR) rate, i.e., the net number of interband energy transitions given by the electrons relaxed into the valence band (recombination rate R_n) minus the electrons promoted into the conduction band (generation rate G_n) – or viceversa for the holes – per unit volume per second. Then, if the BTE for electrons and holes are combined with Eq. (3.4), the following system is obtained

$$\frac{\partial n}{\partial t} = \frac{1}{q}\frac{\partial J_n}{\partial x} - (R_n - G_n)$$
$$\frac{\partial p}{\partial t} = -\frac{1}{q}\frac{\partial J_p}{\partial x} - (R_p - G_p)$$ (3.21)
$$\frac{\partial \mathscr{E}}{\partial x} = \frac{\rho}{\varepsilon}$$

which represents the so-called Drift-Diffusion (DD) model (written for clarity in one space dimension). The first two equations, deriving from the standard BTE, are the electron-hole continuity equations, while the last one is the Poisson's equation.

It is interesting to highlight that in the so-called *lifetime approximation*, equivalent to the relaxation time approximation for the BTE collision term, it is possible to write

$$U_n = R_n - G_n$$
$$\approx \frac{n - n_0}{\tau_n} = \frac{n'}{\tau_n} \tag{3.22}$$

and

$$U_p = R_p - G_p$$
$$\approx \frac{p - p_0}{\tau_p} = \frac{p'}{\tau_p}, \tag{3.23}$$

where n_0 and p_0 are the carrier densities at equilibrium, n' and p' the excess carrier densities (out of equilibrium) and $\tau_{n,p}$ the (doping- and temperature-dependent) recombination lifetimes, that change according to the semiconductor material and to the particular GR process considered. These assumptions – and, thus, the entire DD model – are acceptable only if the system dynamics is sufficiently slower than the lifetimes $\tau_{n,p}$, as always occurs in traditional semiconductors devices. A more detailed description of the possible GR mechanisms in silicon sensors will be provided in the following sections of this chapter.

3.3 NUMERICAL TREATMENT OF MODELS

The equation of the DD model needs to be modified in order to be implemented in a computer program. The appropriate tool to solve a transport equation is the partial differential equation (PDE). The complexity of the system in Eq. (3.21), two nonlinear and one linear equation in four variables, has to be reduced. The usual strategy is to discretize the PDE both in the space and time domains, and transform them into ordinary differential equations (ODE). To achieve this target, the device geometry is divided into a grid of nodes and the dynamic transitions are treated as a sequence of quasi-stationary states. These transitions can be solved with the discretized version of the DD equations, i.e., in each node of the grid, and where all physical quantities are expressed as functions of their nodal values.

At the end of this section, properties of the solving method used to calculate the discretized DD model will be shown, providing information about the possible numerical issues.

3.3.1 METHODS OF SPATIAL DISCRETIZATION

The schemes commonly used to discretize the system are essentially two: the Finite Element (FE) and Finite Differences (FD). To explore the main differences between the FE and FD schemes, the Poisson's equation is used as an example. In its generic form, the Poisson's equation is written as

$$\nabla_{\mathbf{r}}^2 \varphi = f(\mathbf{r}), \tag{3.24}$$

where $f(\mathbf{r})$ is the known term of this PDE. The potential needs to be expressed as a set of basis functions defined in each j-th element of the space such that

$$\varphi \sim \sum_j \phi_j w_j(\mathbf{r}), \qquad (3.25)$$

where ϕ_j is the nodal value of the potential and $w_j(\mathbf{r})$ are opportune weighting functions. So, the Poisson's equation can be written as

$$\sum_j \phi_j \nabla_\mathbf{r}^2 w_j(\mathbf{r}) = f(\mathbf{r}). \qquad (3.26)$$

Now, Eq. (3.26) can be integrated on a given volume of space, called Ω, obtaining

$$\sum_j \phi_j \int_\Omega w_k(\mathbf{r}) \nabla_\mathbf{r}^2 w_j(\mathbf{r}) \, d\mathbf{r} = \int_\Omega w_k(\mathbf{r}) f(\mathbf{r}) \, d\mathbf{r}. \qquad (3.27)$$

This is a matrix equation in the form

$$\mathbf{A} \cdot \mathbf{\Phi} = \mathbf{f}, \qquad (3.28)$$

where

$$A_{jk} = \int_\Omega w_k(\mathbf{r}) \nabla_\mathbf{r}^2 w_j(\mathbf{r}) \, d\mathbf{r} \qquad (3.29)$$

are the elements of the sparse matrix $\mathbf{A}$ and where $\mathbf{\Phi}$ and $\mathbf{f}$ are the column vector of, respectively, the discretized potential and the known term. Now, assuming that Ω is the triangular region defined by three nearby nodes of the grid, Eq. (3.28) becomes the discretized Poisson's equation within the FE scheme, where the unit-elements are the triangular control regions.

Similarly, the grid (with a certain criteria) can be divided into boxes with area S_j that are surrounding each j-th node. If γ_j is the path around the box, the Gauss theorem can be exploited to write Eq. (3.24) as

$$\int_{S_j} \nabla \cdot (\nabla \varphi) \, d\mathbf{r} = \oint_{\gamma_j} \nabla \varphi \cdot \mathbf{n} \, d\ell = \int_{S_j} f(\mathbf{r}) \, d\mathbf{r}. \qquad (3.30)$$

By applying the 1st-order Taylor expansion, the following expression is obtained

$$\oint_{\gamma_j} \frac{\partial \varphi}{\partial \mathbf{n}} \, d\ell \simeq \sum_{\text{sides}} \ell_i \frac{\varphi_j - \varphi_i}{d_{ij}} = S_j f(r_j), \qquad (3.31)$$

where ℓ_i denotes the length of the side of the box around the jth-node, located between the adjacent nodes j and i, d_{ij} is the distance between these two nodes, S_j the area of the jth-box and $\mathbf{n}$ is a unit vector normal to the box side. Notice that Eq. (3.31) is, as in the previous case, a matrix equation

$$\mathbf{A} \cdot \mathbf{\Phi} = \mathbf{S} \cdot \mathbf{f}, \qquad (3.32)$$

where $\mathbf{S}$ is a diagonal matrix. The relation in (3.32) is the discretized version of the Poisson's equation written according to the FD scheme (that, in 2D, is usually called Finite Boxes (FB) method). The core of this scheme, and the reason for its name, is in the term $\varphi_j - \varphi_i$, which takes into account the potential difference between two adjacent nodes.

(a)   (b)

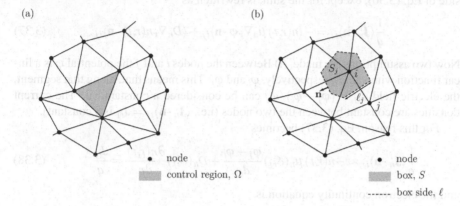

* node
control region, Ω

* node
box, S
- - - - - - box side, ℓ

Figure 3.1 Bidimensional representation of *left*, the Finite Element (FE) and *right*, Finite Boxes (FB) discretization schemes. In the first case, the characteristic element (control region) to which all the physical quantities refer to, is the triangle described by three nearby nodes, whereas in the latter case is the box around a node.

As it was done for the Poisson's equation, the FB method can be exploited to discretize also the continuity equations. Before applying the Gauss theorem, one has to note that both the continuity equations included in the DD model have the form

$$\frac{\partial f}{\partial t} + \nabla \cdot \mathbf{F} = s,$$
(3.33)

where f and $\mathbf{F}$ are, respectively, a scalar and a vector unknown and where s is a (scalar) source term.

The domain of the box, having area S and perimeter Γ, can now be integrated over. The Gauss theorem applied to Eq. (3.33) yields to

$$\frac{\partial}{\partial t} \int_{S_j} f \, dS + \oint_{\Gamma_j} \mathbf{F} \cdot \mathbf{n} \, d\Gamma = \int_{S_j} s \, dS,$$
(3.34)

that becomes

$$\frac{df_j(t)}{dt} S_j + \sum_{\text{sides}} \ell_i (\mathbf{F} \cdot \mathbf{n})_i \approx s_j S_j.$$
(3.35)

The (3.35) applied to the electron continuity equation gives

$$\frac{dn_j(t)}{dt} S_j - \frac{1}{q} \sum_i \ell_{ij} \langle \mathbf{J}_n \cdot \mathbf{n} \rangle_{ij} = -U_{n,j} S_j,$$
(3.36)

where i is a index to identify all the possible box sides around the node j. Most of the terms appearing in Eq. (3.36) can quite easily be managed. What may induce ambiguity is the mean value of $\langle \mathbf{J}_n \cdot \mathbf{n} \rangle$ carried out over all the neighboring nodes. To proceed further, the twofold nature of the current density (drift and diffusion components) needs to be considered. In this way, the second term on the left-hand side of Eq. (3.36), except for the sum, is rewritten as

$$\frac{1}{q} \langle \mathbf{J}_n \cdot \mathbf{n} \rangle_{ij} = -\langle n(r,t) \, \mu_n \nabla_\mathbf{r} \varphi \cdot \mathbf{n} \rangle_{ij} + \langle D_n \nabla_\mathbf{r} n(r,t) \cdot \mathbf{n} \rangle_{ij}. \qquad (3.37)$$

Now two assumptions are made. (i) Between the nodes i and j the potential φ is a linear function with values, respectively, φ_i and φ_j. This means that, along that segment, the electric field $\mathscr{E}_{ij} = (\varphi_j - \varphi_i)/d_{ij}$ can be considered a constant. (ii) The current densities are constant between the two nodes (i.e., $\langle \mathbf{J}_n \cdot \mathbf{n} \rangle_{ij} = J_{ij}$ is a constant).

For this reason, Eq. (3.37) becomes

$$\frac{1}{q} \langle \mathbf{J}_n \cdot \mathbf{n} \rangle_{ij} \approx -n(r,t) \, \mu_n(\mathscr{E}_{ij}) \frac{\varphi_j - \varphi_i}{d_{ij}} + D_n(\mathscr{E}_{ij}) \frac{\partial n(r,t)}{\partial r} = \frac{J_{ij}}{q}, \qquad (3.38)$$

and the electron continuity equation is

$$\frac{dn_j(t)}{dt} S_j + \sum_i \ell_{ij} n(r,t) \, \mu_n(\mathscr{E}_{ij}) \frac{\varphi_j - \varphi_i}{d_{ij}} - \sum_i \ell_{ij} D_n(\mathscr{E}_{ij}) \frac{\partial n(r,t)}{\partial r} = -U_{n,j} S_j, \quad (3.39)$$

with the unknowns n, φ and $\mathscr{E}$ (a dual equation also holds for holes).

Since the solution of Eq. (3.39) may generate stability issues, several approaches have been developed in the past years. One of the most robust and suitable methods (but not the only one) for the implementation in a software program is the so-called Scharfetter-Gummel (SG) solving scheme. Being the stability mainly due to the behaviour of the function $n(r,t)$ (as well as of the hole density), the SG approach makes some particular assumptions on the carrier density and its trend between the nodes that allows reducing the continuity equations to the form

$$\frac{dn_j(t)}{dt} \overset{\text{SG}}{\approx} \sum_i \frac{D_n(\mathscr{E}_{ij})}{\ell_{ij} S_j} \left[n_j(t) \, \mathsf{B}(\Delta\varphi) - n_i(t) \, \mathsf{B}(-\Delta\varphi) \right] - U_{n,j}$$
$$\frac{dp_j(t)}{dt} \overset{\text{SG}}{\approx} -\sum_i \frac{D_p(\mathscr{E}_{ij})}{\ell_{ij} S_j} \left[p_j(t) \, \mathsf{B}(\Delta\varphi) - p_i(t) \, \mathsf{B}(-\Delta\varphi) \right] - U_{p,j} \qquad (3.40)$$

where

$$\mathsf{B}(\Delta\varphi) = \frac{\Delta\varphi}{\exp(\Delta\varphi) - 1} \qquad (3.41)$$

is a Bernoulli function and where $\Delta\varphi \equiv \varphi_j - \varphi_i$.

At the end of the calculations, a formalism that allows converting all the continuous physical quantities into nodal values, which depend on the discretization scheme chosen to simplify the system, has been obtained. Equation (3.40) differs from its original form since the PDEs are now a set of ODEs. As stated in the introduction of the present chapter, this represents the ideal scenario for the implementation of the transport model into a numerical solver.

3.3.2 THE ITERATIVE SOLUTION OF THE EQUATIONS

A physical model described in analytical form cannot be handled easily by a numerical solver: it needs to be rewritten in a discretized formalism. This paragraph shows this procedure for the Poisson's equation, while leaving the treatment of the whole DD model to the reader's interest. The simplest expression of the Poisson's equation is written as

$$\nabla^2 \varphi(\mathbf{r}) = -\frac{q}{\varepsilon} \left[n(\varphi(\mathbf{r})) + N_A(\varphi(\mathbf{r})) - p(\varphi(\mathbf{r})) - N_D(\varphi(\mathbf{r})) \right]. \qquad (3.42)$$

Two assumptions are necessary: (*i*) the effective densities of donors and acceptors (N_D and N_A) are kept constant with respect to the electrostatic potential (and coordinate $\mathbf{r}$); (*ii*) the potential can be written as

$$q \varphi(\mathbf{r}) \equiv u(\mathbf{r}), \qquad (3.43)$$

where $u(\mathbf{r})$ is a given analytical function.

It follows that, in one space dimension x, Eq. (3.42) becomes

$$\frac{d}{dx} \left(\varepsilon \frac{du(x)}{dx} \right) = q^2 \left[N_D - N_A + p(u(x)) - n(u(x)) \right]. \qquad (3.44)$$

Suppose now to evaluate Eq. (3.44) for the unknown $u(x)$ only in a set of equally spaced $(N+1)$ points x_i, where $i = 0, \ldots, N$. First, one has to impose a solution at the edges of the domain x_0 and x_N. These values are the boundary conditions (BC) of the problem. With the BC set, the solution is restricted to only $(N-1)$ points. Then, the Poisson's equation is evaluated in a generic node x_i through the finite difference scheme, where the interval $\left[x_{i-1/2}; x_{i+1/2} \right]$ was used as a control region to perform the calculations. From what has been said, in x_i it holds that

$$\int_{x_{i-\frac{1}{2}}}^{x_{i+\frac{1}{2}}} \frac{d}{dx} \left(\varepsilon \frac{du(x)}{dx} \right) dx = q^2 \int_{x_{i-\frac{1}{2}}}^{x_{i+\frac{1}{2}}} \left[N_D - N_A + p(u(x)) - n(u(x)) \right] dx. \qquad (3.45)$$

Thanks to the properties of definite integrals it follows

$$\int_{x_{i-\frac{1}{2}}}^{x_{i+\frac{1}{2}}} \frac{d}{dx} \left(\varepsilon \frac{du(x)}{dx} \right) dx = \varepsilon_{x_{i+\frac{1}{2}}} \frac{du(x)}{dx} \bigg|_{x_{i+\frac{1}{2}}} - \varepsilon_{x_{i-\frac{1}{2}}} \frac{du(x)}{dx} \bigg|_{x_{i-\frac{1}{2}}} \qquad (3.46)$$

$$= \varepsilon_{x_{i+\frac{1}{2}}} \frac{u(x)(x_{i+1}) - u(x_i)}{\Delta x} - \varepsilon_{x_{i-\frac{1}{2}}} \frac{u(x_i) - u(x_{i-1})}{\Delta x},$$

where, in the second line, the definition of difference quotient has been used to evaluate the two derivatives. Assuming now that N is sufficiently high to have a dense set of nodes, then $u(x)$ can be approximated by a linear function around the point x_i and, more in general, within the whole control volume.

This hypothesis leads to

$$q^2 \int_{x_{i-\frac{1}{2}}}^{x_{i+\frac{1}{2}}} [N_D - N_A + p - n] \approx q^2 [N_D - N_A + p - n]_{x_i} \Delta x, \qquad (3.47)$$

where Δx is defined as the spacing between nodes (which is constant over the entire domain x). Combining equations (3.46) and (3.47), a new discretized version of the Poisson's equation is obtained:

$$\varepsilon_{x_{i+\frac{1}{2}}} u(x_{i+1}) - \left(\varepsilon_{x_{i+\frac{1}{2}}} + \varepsilon_{x_{i-\frac{1}{2}}} u(x_i) \right) + \varepsilon_{x_{i-\frac{1}{2}}} u(x_{i-1})$$
$$- (\Delta x)^2 \, q^2 \, [N_D - N_A + p\,(u(x_i)) - n\,(u(x_i))] = 0, \qquad (3.48)$$

which has three unknowns: $u(x_{i-1})$, $u(x_i)$, and $u(x_{i+1})$.

The expression obtained is non-linear with respect to the unknowns since the charge carriers n and p are, in turn, non-linear in the potential term. To solve this equation in all the nodes, a numerical strategy that overcomes such an issue needs to be applied. One of the most used formalism is the iterative Newton's method. Besides the BC, the Newton's method also requires opportune Initial conditions (IC) of the system. In this case, the *charge neutrality law* at equilibrium, consisting in

$$N_D - N_A + p\,(u(x)) - n\,(u(x)) = 0, \qquad (3.49)$$

can be chosen for this aim. Simplifying, for each node x_i an equation of the form

$$f_i(u_{i-1}, u_i, u_{i+1}) = 0, \qquad (3.50)$$

is found, where the notation has been relaxed such that now $u_i \equiv u(x_i)$. The goal of the Newton's method is to provide an approximate solution of Eq. (3.50) starting from the IC and through subsequent iterations k, each one having a guess solution to be achieved within a certain tolerance. The iterative method requires a maximum precision $(\Delta u)_{\max}$ as input parameter (automatic or user-defined) and assumes, for all nodes x_i, that

$$u^k = u^{k-1} + \Delta u^k, \qquad (3.51)$$

with u^k the solution at the kth-iteration and Δu^k the difference between two consecutive outcomes. Here, Δu^k represents the progressive correction factor of the method towards the final solution.

By applying the scheme (3.51) to the Eq. (3.50), the following expression is obtained:

$$f_i(u_{i-1}^k, u_i^k, u_{i+1}^k) = f_i \left(u_{i-1}^{k-1} + \Delta u_{i-1}^k, u_i^{k-1} + \Delta u_i^k, u_{i+1}^{k-1} + \Delta u_{i+1}^k \right) = 0, \qquad (3.52)$$

which is, finally, a system of linear equations in u where each solution at k depends on the solution found at the $(k-1)$th iteration. At the step $k = 1$ the value u^1 is a function of u^0, the so-called *initial guess* of the iterative scheme. Each solver estimates this term through different techniques, depending on the application field. The final form of the Poisson's equation is obtained by rewriting Eq. (3.52) as a first-order Taylor expansion:

$$f_i\left(u_{i-1}^{k-1}+\Delta u_{i-1}^k, u_i^{k-1}+\Delta u_i^k, u_{i+1}^{k-1}+\Delta u_{i+1}^k\right) \approx f_i(u_{i-1}^k, u_i^k, u_{i+1}^k)+$$

$$+ \left.\frac{\partial f_i}{\partial u_{i-1}}\right|_{u_{i-1}^k}\Delta u_{i-1}^k + \left.\frac{\partial f_i}{\partial u_i}\right|_{u_i^k}\Delta u_i^k + \left.\frac{\partial f_i}{\partial u_{i+1}}\right|_{u_{i+1}^k}\Delta u_{i+1}^k = 0. \quad (3.53)$$

Equation (3.53) can be cast into matrix form:

$$
\begin{pmatrix}
\bullet & \bullet & & & & 0 \\
\bullet & \bullet & \bullet & & & \\
& \bullet & \bullet & \bullet & & \\
& & \ddots & \ddots & \ddots & \\
& & & \bullet & \bullet & \bullet \\
& & & & \bullet & \bullet \\
0 & & & & \bullet & \bullet
\end{pmatrix}
\begin{pmatrix} \Delta \mathbf{u}^k \end{pmatrix}
= -\begin{pmatrix} \mathbf{f} \end{pmatrix}
\quad (3.54)
$$

which is composed by a tri-diagonal matrix and the column vectors $\Delta \mathbf{u}^k$ and $\mathbf{f}$, that are, respectively, the correction and the residual vector.

The iterative method proceeds until at least one of the following requirements is satisfied: $\|\Delta \mathbf{u}^k\| < \delta$ or $\|\mathbf{f}\| < \delta$, where $\delta \equiv (\Delta u)_{max}$ is the tolerance of the Newton's method. For any given x_i and k, if a solution is found within a finite number of iterations, then the method converges, otherwise the procedure does not converge.

Reason determining a non-converging system are (not exhaustive): (*i*) inadequate boundary conditions, (*ii*) poor discretization scheme (extremely dense or coarse mesh nodes), (*iii*) too small tolerance, and (*iv*) low computational power or (*v*) badly conditioned problems (for instance, due to a high number of charges or a large domain to be simulated).

In order to minimize the risk that non-convergence occurs, the Russian mathematician B. N. Delaunay developed in 1934 a robust triangulation procedure that makes use of non-obtuse triangles [13]. In combination with a particular domain tessellation that identifies the finite boxes by connecting the three bisector lines of each Delaunay triangle, a discretization scheme providing an even more stable solution, particularly suitable for simulating semiconductor devices, is found.

3.4   UFSD IMPLEMENTATION AND MODELLING

The aim of this section is to present the simulation of a UFSD-based particle detector and, at the same time, to show a real application of the concepts exposed in the previous paragraphs. Given the complexity of a silicon detector, only a model based on a numerical approach provides a reliable description. In this regard, the Technology

Computer-Aided Design (TCAD) is the most used solution. The TCAD implementation of the DD model requires a set of mathematical handles in order to solve the
equations through the iterative method, for example the tolerance and the maximum
number of iterations. In addition, for any structure to be modelled, all the geometrical and physical properties have to be taken into account. The materials and all the
topological elements necessary to correctly impose the BC (as in the case of contacts
or external edges), need to be defined. This step includes the definition of doping
implants and profiles, and the presence of structural defects in the lattice having relevance in the physics of carrier transport, such as energy traps or fixed charges due to
some fabrication processes. TCAD also allows simulating ion implantation, thermal
annealing, crystal growth or material deposition. These features can provide more
realistic scenarios in the event that such technological details can make a difference.
Besides the electromagnetic and transport equations just derived, it is necessary to
activate in the simulations the models of all physics processes deemed essential to
the operation of the device. These are the processes that can alter the current flow in
the device under test and, for such reason, must be implemented into the DD model
as appropriate GR terms.

So far, only the core of the numerical modelling has been described. The following step is to define the target of the simulation (for example a voltage ramp or
transient phenomena). The space domain is divided into a grid and the discretized
version of the DD equations is solved in each node of the grid, where all the physical
quantities are expressed as functions of their nodal values. A solid procedure is to use
variable node spacing: where the relevant quantities or their gradients are expected to
be particularly high, a finer mesh is necessary to allow the numerical solving. In case
of transients, the time domain has to be discretized too so that a sweep of a physical
quantity or the simulation of a time-dependent process is treated as a sequence of
quasi-stationary states. As for the space domain, the density of steps must properly
follow the time scale of the process: the quasi-stationary approximation is based on
the assumption that the interval between two steps must be significantly shorter than
the time of the transient.

Once the simulation domains are set up, the user selects the tolerance, the maximum number of iterations as well as all the BC and IC of the problem. Then the tool
introduces in each node the initial guess for the Poisson's equation at equilibrium
($t = 0$), which is the electrostatic potential φ_0'. After solving the Poisson's equation,
a numerical estimation φ_0 of the potential is obtained. This solution can be used as
the initial guess to solve the whole DD system in the first time step, $t = 1$. Using the
iterative scheme, TCAD estimates all the DD unknowns (charge densities and potential and, in turn, also the electric field and current densities), until the simulation
goal is achieved. This happens only if the algorithm converges for each time step.

3.4.1 GENERATION-RECOMBINATION MECHANISMS

The physical processes occurring in silicon sensors are mostly defined in terms of a
GR rate, representing the net number of generated or recombined electrons/holes,
depending on which of the two competing processes is dominating. Before

introducing the physics used to model the UFSD-based silicon detectors, the minimum settings of any simulation is briefly discussed here.

First, all the energy transitions between the valence and conduction band that are assisted by those lattice impurities (like defects of dopants) acting as energy traps have to be considered. This kind of processes goes under the name of Shockley-Read-Hall (SRH) GR mechanisms [120], from the name of the scientists who developed this formalism. Out of equilibrium, the net GR rates $U_{n,p}$ are the algebraic sum of the terms $(R_{n,p} - G_{n,p})$. The statistical nature of the emission and capture processes is defined by introducing the *capture* and *emission coefficients* for electrons and holes $c_{n,p}$ and $e_{n,p}$. Both terms are a constant property (in s^{-1}) of a given trapping process. Thus, the SRH recombination rates (in cm^{-3}s^{-1}) is defined as

$$R_n = c_n N_t \left(1 - f(E_t)\right) n$$
$$R_p = c_p N_t f(E_t) p \tag{3.55}$$

where N_t is the total number of SRH trap states, E_t its energy, $f(E_t)$ the probability that a trap with energy E_t is occupied, and $1 - f(E_t)$ the probability to find that trap unoccupied.

Symmetrically, the generation rates (in cm^{-3}s^{-1}) are

$$G_n = e_n N_t f(E_t)$$
$$G_p = e_p N_t \left(1 - f(E_t)\right) \tag{3.56}$$

being the recombination a capture process of an electron coming from the conduction band and the generation an emission that excites an electron from the trap state to the conduction band (see Fig. 3.2). It can be demonstrated that, in steady-state conditions, the electron and hole net rates are equal, giving

$$c_n N_t \left[(1 - f(E_t)) n - f(E_t) n_0 (E_F)\right] = c_p N_t \left[f(E_t) p - (1 - f(E_t)) p_0 (E_F)\right] \tag{3.57}$$

where $n_0 (E_F)$ and $p_0 (E_F)$ are the equilibrium electron-hole concentrations when the trap has energy $E_t \equiv E_F$ (with E_F the Fermi level). The solution to this equation is

$$f(E_t) = \frac{c_n n + c_p p_0 (E_F)}{c_n (n + n_0 (E_F)) + c_p (p + p_0 (E_F))}. \tag{3.58}$$

Combining equations (3.57) and (3.58), the net rates are obtained:

$$U_n = U_p = \frac{np - n_i^2}{\tau_p (n + n_0 (E_F)) + \tau_n (p + p_0 (E_F))}, \tag{3.59}$$

where the *mass action law*

$$np = n_0 (E_F) p_0 (E_F) = n_i^2 \tag{3.60}$$

was used, being n_i the intrinsic carrier concentration, and with the assumption that

$$\tau_{n,p} = \frac{1}{c_{n,p} N_t} \tag{3.61}$$

The formula written in Eq. (3.59) represents the most used expression of the net SRH generation-recombination rate. In order to be included in the DD problem, it must be plugged into the right-hand side of both continuity equations and then self-consistently solved with the Poisson's equation.

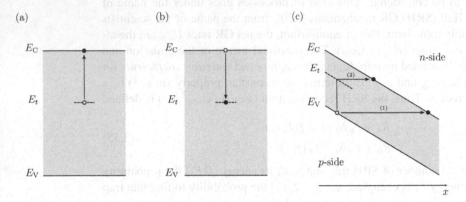

Figure 3.2 Band diagrams showing three different processes: (a) electron emission from a trap state into the conduction band (generation); (b) electron capture (recombination); (c) band-to-band (1) and trap-assisted (2) tunneling (generation) mechanisms in a reversely biased *pn* junction. Labels E_C, E_V, and E_t represent the conduction band, valence band, and trap energy, respectively.

The other important family of GR processes is represented by the tunneling mechanisms. They differ from the SRH ones by the fact that tunneling is a transition in space and not (necessarily) in energy. Again, trap-assisted tunneling processes (TAT) are found, starting or finishing with a trap capture/emission, or direct band-to-band tunneling processes (BTBT), taking place without any intermediate level. They may also occur in UFSD-based detectors, when the band bending far from the equilibrium determines a distance encompassing the conduction and valence band edges which is comparable to the wavelength of carriers, or when the presence of traps is such that this path is physically reduced, even with lower applied field. For this kind of mechanisms, net rates have the form [88, 89]

$$U_{\text{TAT}} = \frac{np - n_i^2}{\tau_p(\mathscr{E})\left(n + n_i\, e^{\frac{E_t - E_{\text{F},i}}{k_\text{B}T}}\right) + \tau_n(\mathscr{E})\left(p + p_i\, e^{\frac{E_{\text{F},i} - E_t}{k_\text{B}T}}\right)} \tag{3.62}$$

and [93]

$$U_{\text{BTBT}} = A\,\mathscr{E}^2\, e^{-B/\mathscr{E}}, \tag{3.63}$$

where $E_{\text{F},i}$ is the energy of the intrinsic Fermi level (at the mid-gap), and the coefficients A and B are material-dependent constants. Both these two latter expressions represent theoretical models deriving from *ab initio* calculations based on, respectively, the multiphonon emission theory and on the $\mathbf{k} \times \mathbf{p}$ approximation of bands [99].

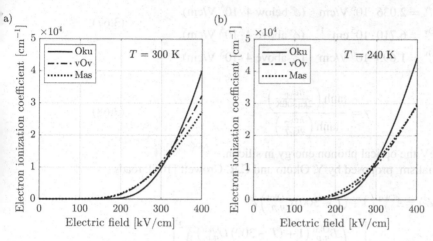

Figure 3.3 Trend of the electron ionization coefficient α_n as a function of the applied field, according to the three different avalanche models and for two temperatures: (a) at $T = 300$ K and (b) $T = 240$ K.

What characterizes the simulation of a UFSD device is the presence of the gain layer. This implies, from the numerical standpoint, that a model describing the avalanche of charge multiplication needs to be implemented. All the avalanche processes usually have a net rate of the form

$$U_{\text{aval}} = \alpha_n(\mathscr{E})\, n\, v_n + \alpha_p(\mathscr{E})\, p\, v_p \,, \tag{3.64}$$

in which $v_{n,p}$ are the carrier drift velocities and $\alpha_{n,p}(\mathscr{E})$ are the so-called electron-hole *ionization coefficients* (in cm^{-1}), corresponding to the inverse of the mean free path between two subsequent scattering events producing secondary charges. The various models developed in the past years differ in the form of the ionization coefficients. Three formalisms, that follow the Chynoweth law [7], are normally employed. The first one is the van Overstraeten-de Man model [133]:

$$\alpha_{n,p}^{\text{vOv}}(\mathscr{E}) = \gamma A_{n,p}^{\text{vOv}} \exp\left(-\gamma \frac{B_{n,p}^{\text{vOv}}}{\mathscr{E}}\right), \tag{3.65}$$

whose parameters are

$$\begin{aligned} A_n^{\text{vOv}} &= 7.030 \cdot 10^5 \text{ cm}^{-1} \\ B_n^{\text{vOv}} &= 1.231 \cdot 10^6 \text{ V/cm} \,, \end{aligned} \tag{3.66}$$

$$A_p^{\text{vOv}} = 1.582 \cdot 10^6 \, \text{cm}^{-1} \quad (\mathscr{E} \text{ below } 4 \cdot 10^5 \, \text{V/cm})$$

$$B_p^{\text{vOv}} = 2.036 \cdot 10^6 \, \text{V/cm} \quad (\mathscr{E} \text{ below } 4 \cdot 10^5 \, \text{V/cm})$$

$$A_p^{\text{vOv}} = 6.710 \cdot 10^5 \, \text{cm}^{-1} \quad (\mathscr{E} \text{ above } 4 \cdot 10^5 \, \text{V/cm})$$

$$B_p^{\text{vOv}} = 1.693 \cdot 10^6 \, \text{V/cm} \quad (\mathscr{E} \text{ above } 4 \cdot 10^5 \, \text{V/cm})$$

(3.67)

and

$$\gamma = \frac{\tanh\left(\frac{\hbar\omega_{\text{op}}}{2k_{\text{B}}300\text{K}}\right)}{\tanh\left(\frac{\hbar\omega_{\text{op}}}{2k_{\text{B}}T}\right)},$$

(3.68)

with $\hbar\omega_{\text{op}} = 0.063$ eV the optical phonon energy in silicon.

The second formalism, proposed by Y. Okuto and C.R. Crowell [105], reads

$$\alpha_{n,p}^{\text{Oku}}(\mathscr{E}) = A_{n,p}^{\text{Oku}} \left(1 + (T - 300) C_{n,p}^{\text{Oku}}\right) \mathscr{E}$$

$$\times \exp\left[-\left(\frac{B_{n,p}^{\text{Oku}} \left(1 + (T - 300) D_{n,p}^{\text{Oku}}\right)}{\mathscr{E}}\right)^2\right],$$

where

$$A_n^{\text{Oku}} = 0.426 \, \text{V}^{-1}$$

$$A_p^{\text{Oku}} = 0.243 \, \text{V}^{-1}$$

$$B_n^{\text{Oku}} = 4.81 \cdot 10^5 \, \text{V/cm}$$

$$B_p^{\text{Oku}} = 6.53 \cdot 10^5 \, \text{V/cm}$$

(3.69)

and

$$C_n^{\text{Oku}} = 3.05 \cdot 10^{-4} \, \text{K}^{-1}$$

$$C_p^{\text{Oku}} = 5.35 \cdot 10^{-4} \, \text{K}^{-1}$$

$$D_n^{\text{Oku}} = 6.86 \cdot 10^{-4} \, \text{K}^{-1}$$

$$D_p^{\text{Oku}} = 5.67 \cdot 10^{-4} \, \text{K}^{-1}.$$

(3.70)

Finally, in the Massey model [100], the ionization coefficients are written as

$$\alpha_{n,p}^{\text{Mas}}(\mathscr{E}) = A_{n,p}^{\text{Mas}} \exp\left(-\frac{B_{n,p}^{\text{Mas}}(T)}{\mathscr{E}}\right),$$

(3.71)

with parameters

$$A_n^{\text{Mas}} = 4.43 \cdot 10^5 \, \text{cm}^{-1}$$

$$A_p^{\text{Mas}} = 1.13 \cdot 10^6 \, \text{cm}^{-1}$$

(3.72)

$$B_n^{\text{Mas}}(T) = C_n^{\text{Mas}} + D_n^{\text{Mas}} \cdot T$$

$$B_p^{\text{Mas}}(T) = C_p^{\text{Mas}} + D_p^{\text{Mas}} \cdot T,$$

(3.73)

$$C_n^{\text{Mas}} = 9.66 \cdot 10^5 \text{ V/cm}$$

$$C_p^{\text{Mas}} = 1.71 \cdot 10^6 \text{ V/cm},$$

(3.74)

and

$$D_n^{\text{Mas}} = 4.99 \cdot 10^2 \text{ V cm}^{-1} \text{ K}^{-1}$$

$$D_p^{\text{Mas}} = 1.09 \cdot 10^3 \text{ V cm}^{-1} \text{ K}^{-1}.$$

(3.75)

The Okuto-Crowell and Massey models have a more pronounced dependence of the ionization coefficients on temperature with respect to the van Overstraeten-de Man one, as it can be observed by comparing the two plots of Fig. 3.3.

3.4.2 RADIATION DAMAGE MODELLING

Typically, TCAD tools do not include built-in functions accounting for specific bulk radiation damage models in silicon particle detectors. To overcome this fact, *ad-hoc* models are added in the numerical framework. As seen in Section 1.3, a simple formulation describing the production of acceptor-like defects and the deactivation of acceptor dopants in UFSD-based detectors is [48]

$$N_A(\Phi, x) = g_{\text{eff}} \Phi + N_A(0, x) e^{-\Phi \cdot c(N_A(0, x))},$$

(3.76)

where N_A is the acceptor density in silicon, Φ is the fluence (in $n_{\text{eq}}/\text{cm}^2$), g_{eff} a coefficient determining the effective acceptor states production, and c is an appropriate function of the acceptor density before irradiation $N_A(0, x)$. This function indicates how strongly radiation deactivates the acceptor atoms. Notice that N_A is also a function of the position x inside the device. This means that both the acceptor removal process, described by the last term of Eq. (3.76), and the function c change according to the initial local density.

When performing parametric UFSD simulations where the unknowns are functions of the irradiation level, Eq. (3.76) is implemented in the system in such a way that the variable x maps each node of the discretization grid. To this aim, two steps are required: first of all, the user has to recompute off-line all the p-type profiles included in the detector according to the acceptor removal-creation law. Secondly, this profile is discretized and plugged into the Poisson's and continuity equations so that the initial conditions can include the new doping when solving the DD model. Usually, TCAD programs automatically adapt the profile discretization created by the user to the mesh, which is a remarkable advantage.

A different procedure can be exploited to define radiation-induced trap states in oxides or at interfaces between different materials, where the physical models governing their creation are typically simpler than Eq. (3.76). Since most of the TCAD tools predispose the implementation of defect levels or bands, the user should introduce their characteristic parameters (i.e., the energy and their eventual energy distribution, in case of band states, the initial concentration and the scattering or GR cross-sections with electrons and holes). If the experiments suggest that these defects acting like traps vary with the fluence in terms of density or energy, the user needs to specify the parametric law through which the dependence takes place and TCAD automatically recomputes them material-wise or region-wise by reason of the radiation dose. One of the most robust frameworks in the literature describing how surface damage behaves with fluence is the so-called Perugia model [28, 29], a phenomenological set of parameters which introduces, besides the oxide charges, also acceptor- and donor-like interface trap states. In particular, acceptors uniformly occupy a band included between $E_C - 0.56$ eV and E_C, while donors are distributed in a band of width 0.6 eV, starting in correspondence of the valence band energy E_V. The concentrations of these defects, as well as of the charges in the oxide, are fluence-dependent and may slightly change according to the foundry producing the devices.

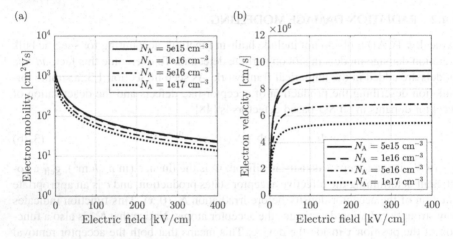

Figure 3.4 Trend of the electron mobility, *left*, and velocity in silicon *right*, as a function of the parallel component (with respect to the drift lines) of the electric field, calculated analytically for four acceptor atoms concentrations N_A, and at room temperature.

When the number of free charges changes due to an effective doping variation, also two fundamental solid-state quantities driving the operation of a silicon device change: the mobility and velocity of carriers (see, for instance, Ref. [3]). This effect is shown, for electrons, in Fig. 3.4, where the variation of both quantities as a function of the electric field has been calculated considering a silicon sample homogeneously doped with four different acceptor concentrations N_A. As one may see, when N_A increases, both quantities decrease for a given field. The curves here reported have a trend well-known in the literature and refer to a specific doping level. This means that they are valid locally, and the global behaviour of the whole device is given by the contribution of all the different doping concentrations. Since the effective doping concentration also depends on the fluence, in order to simulate the impact of radiation damaging on carrier mobility and velocity at the device-level, it is necessary to compute the new physical quantities in each node of the discretization mesh. Moreover, if the acceptor states generated by the radiation are explicitly declared as traps, the solver will also treat them as further scattering centers in which calculate appropriate GR rates for trapping processes. This feature provides a more realistic simulation, especially if the goal is to study in detail the signal shape due to a charged particle passing through the detector. Generally, the radiation has the overall effect of decreasing both mobility and velocity in UFSD-based detectors.

3.4.3 OTHER PHYSICAL MODELS

Due to the crucial role of dopants in the operation of silicon devices, two additional effects, both function of the doping concentration, may be required in UFSD simulations. The first mechanism is the so-called *band-gap narrowing*, consisting – as the name suggests – in a slight reduction of the forbidden energy gap of silicon. This is due to the orbital overlapping of dopant atoms when they exceed a critical density N_{crit}. The energy bands generated by these new states may be sufficiently shallow and wide (the band is larger than E_{crit}) that they enter the conduction or valence band, with the result of narrowing the gap E_g. The gap reduction at $T = 300$ K is of the order of few tens of meV when the dopants are about 10^{18} cm^{-3}, and slightly decreases with temperature. The higher the total doping density N_{tot}, the larger the band-gap narrowing ΔE_g is. One of the most used formalisms is the Slotboom model [95, 123, 124, 125], according to which

$$\Delta E_g = E_{crit} \left(\ln \left(\frac{N_{tot}}{N_{crit}} \right) + \sqrt{\frac{1}{2} + \left[\ln \left(\frac{N_{tot}}{N_{crit}} \right) \right]^2} \right), \qquad (3.77)$$

where, in silicon, $N_{crit} = 1.3 \cdot 10^{17}$ cm^{-3} and $E_{crit} = 6.92 \cdot 10^{-3}$ eV.

So far, we always referred to the dopants without distinguishing between implanted dose and effective doping concentration. The second important phenomenon involving both the doping and the temperature is the capability of the dopant atoms to provide free charges (negative for donors and positive for acceptors). Since the ion implantation mostly generates interstitial impurities, a thermal process – called annealing – is performed after the implantation to activate the dopants, that become substitutional. During the annealing, besides the activation, also the lattice repair takes place, since the ion implantation has locally induced dislocations, vacancies, point defects or stacking faults. As the annealing time or temperature becomes higher, the lattice rearranges and dopants diffuse. Usually, this process can activate only a fraction of the nominal quantity of implanted atoms, which comes from a delicate balance between applying the minimum energy needed to reduce the lattice defects and keeping under control the diffusion of dopants. So, the thermal cycles applied to the implanted sample must follow a precise recipe, where the process temperature and duration have to be accurately determined. Even supposing to activate all the impurities, once the annealing has been performed, also the operating temperature of the device, in principle, can act on the capability to provide free charge carriers. In fact, each dopant is characterized by a ionization energy, given by the energy difference between the impurity level and the corresponding band edge (the bottom of the conduction band for donors or the top of the valence band for acceptors). When the thermal energy of the system – related to the device

operating temperature – is lower than the ionization energy of dopants the resulting fraction of active impurity atoms participating to the conduction is lower than
100%. This effect is known as *incomplete ionization* and, besides the temperature,
depends also on the nominal concentration of dopants as well as on their chemical
nature.

(a)   (b)

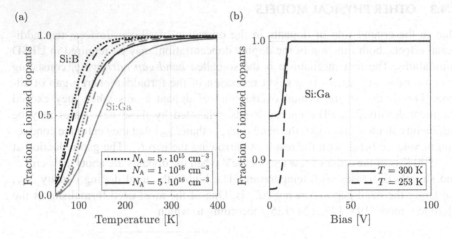

Figure 3.5 *Left*: fraction of activated acceptors in silicon, calculated for boron and gallium at
equilibrium, as a function of the absolute temperature and for three different nominal acceptor
atoms concentrations. *Right*: fraction of activated acceptors, in a UFSD-based structure, with
gallium peak dose $N_A \sim 5 \cdot 10^{16}$ cm^{-3}, plotted at 300 K and 253 K, as a function of the applied
reverse bias.

The *left* panel of Fig. 3.5 shows the fraction of the active acceptor dopants as a
function of the simulation temperature in *p*-type silicon for three different nominal
concentrations of boron and gallium. These calculations have been performed by
using the incomplete ionization law

$$N_{A,0}(\mathscr{E},T) = N_A \left(1 + g_A \exp\left(\frac{E_A - E_{F,p}(\mathscr{E})}{k_B T} \right) \right),$$ (3.78)

which essentially is the Fermi-Dirac distribution for a population of acceptors (similarly for donors) with ionization energy E_A and density of active dopant atoms N_A.
The terms $E_{F,p}$ and g_A are, respectively, the quasi-Fermi level of holes and the degen-
eracy factor, an integer number changing according to the dopant element and equal
to 2 in case of gallium or boron. Equation (3.78) is applied until $N_{A,0}$ is lower than
an effective concentration, above which we usually consider all dopants completely
ionized (in Si:B, such threshold value is 10^{22} cm^{-3}). As the plot shows, the higher
the nominal concentration, the lower the ionization fraction at fixed temperature, for
both boron and gallium atoms.

Since the energy difference $(E_{F,p} - E_V)$ changes with the applied field, it is also
possible to study how the activation evolves out of equilibrium with the external

bias. To this purpose, a UFSD-like structure with a gallium-doped multiplication layer has been implemented in TCAD. The *right* panel, (b), of Fig. 3.5 demonstrates that, for a peak density $\sim 5 \cdot 10^{16}$ cm^{-3} of the gallium implant and two different temperatures, the acceptors are almost completely activated even at low bias, far below the operating point of a standard UFSD-based detector.

3.5 SIMULATING ULTRA-FAST SILICON DETECTORS

In this section, several examples of Ultra-Fast Silicon Detectors simulations, and their comparison with experimental data, will be presented. In the following, if not otherwise specified, the TCAD results are obtained by solving the DD model, which includes SRH generation-recombination, avalanche multiplication, BTBT, TAT, band-gap narrowing and, when necessary, also proper radiation damage models.

The present section is divided into two paragraphs, the first one focuses on the use of numerical simulations to replicate the detector leakage currents and internal electric fields, while the second one on predicting the signal properties in both not irradiated and irradiated detectors.

3.5.1 STATIC CHARACTERISTICS AND ELECTRIC FIELD

As seen when introducing the numerical implementation of physical models, the sweep of an electrical quantity – e.g., the applied bias – corresponds to a sequence of steps, each representing a quasi-stationary state. By solving the DD equations in the whole space domain, it is possible to predict the trend of the total current versus the applied potential, the $I(V)$ characteristics. In order to have realistic results for both the leakage current and the breakdown voltage, the simulations must include as many technological details as possible. Among them, the doping profiles (and also their eventual lateral diffusion spreadings, when simulating in 2D-3D), the intergap defect levels, the interface or oxide charges, and all the geometrical features characterizing the device like the spacing among different implants and structures, or the thickness of each layer.

Figure 3.6 shows in the *left* panel the comparison between the measured and simulated $I(V)$ characteristics of UFSD2 detectors, differing in type and dose of acceptors in the gain implant. The increase of the current is more and more pronounced as the applied reverse voltage is raised due to the effect of charge multiplication. This is a clear example of why a correct modelling of the avalanche mechanism is so crucial in UFSD-based detectors. The *right* panel, instead, shows two $C(V)$ characteristics of two devices from the same UFSD production. The numerical calculation matches well the experimental data and can accurately predict the depletion voltage of both the gain layer and the active substrate. In these calculations, the van Overstraeten–de Man model has been adopted and a quasi-1D simulation domain has been used, having implemented just a silicon slice including the UFSD junction. It is worth noting that this kind of numerical characterizations can be used not only to adapt and calibrate the models and their parameters with respect to the specific

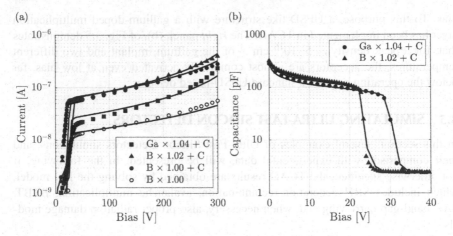

Figure 3.6 *Left*: measured and simulated current-voltage $I(V)$ characteristics. *Right*: measured and simulated capacitance-voltage $C(V)$ characteristics. The curves are for several UFSD2 samples having different dose (from 1.00 to 1.04) and species (gallium and boron) of acceptors, with or without the co-implantation of carbon atoms. Symbols are laboratory measurements, while lines are TCAD simulations [51, 54].

fabrication technology, but also – and especially – to predict the behaviour of the detectors or, in other words, to design them. In this regard, it should be highlighted that the $I(V)$ and $C(V)$ characteristics can help in defining the optimal gain implant profile, for example through the study of breakdown voltages. The same can be said for the gain curves, analyzed in the next section. Besides ensuring that the internal field is well distributed at the periphery, the designer has to assure electrical isolation between nearby active regions (gain implants) in correspondence of the inter-pad regions.

Through a 2D or 3D implementation of the UFSD under study, it is possible to infer important conclusions about the trend of the field and drift lines. Figure 3.7 shows a simulated cross-sectional view of the inter-pad separating two adjacent active regions (the p-gain implants). On both sides, there is a n-type implant, the junction termination extension (JTE), while in the middle a p-stop structure is implanted. The JTEs are used to confine the high fields produced in reverse bias by the gain layer, and to prevent particles crossing the detector in the inter-pad from generating out-of-time signals (see Section 2.8). The p-stop implants, instead, avoid that the electrons inversion layer, due to the oxides and interface charges, short-circuits two adjacent pads. These isolation structures determine – as it will become more clear in the next paragraph – a performance drop because where there is no multiplication the detector gain tends to be close to one. The calculated drift lines help in determining the width of the no-gain region (see, for instance, the lines marked in white), since they allow to predict which is the volume of silicon affected by the lower charge collection. By properly tuning the technological parameters of the isolation implants (such as dose and depth) it is possible to minimize their impact on the detector performance.

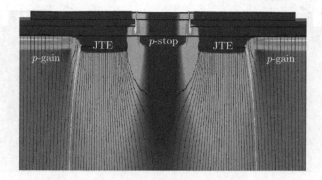

Figure 3.7 Cross-section of a generic UFSD device in the inter-pad region between two neighboring active areas, showing the electron drift lines and the electric field map (in grey tones, where brighter areas have a higher field intensity). White lines are the external drift lines enclosing the JTE implants. They highlight the collection volume of the p-gain.

Other considerations may also concern the chemical nature and thickness of the oxide and passivation layers deposited on the top of the detector that, as well as the use of proper interface defects, represents an important feature, especially for irradiated devices.

Besides the implantation properties, simulations are also used to understand the effects on the internal field distribution of two important key-elements: the oxide thickness and the metal extensions on the device surface. Both concur in determining the device operation, either of the detector periphery, the cut-line region at the physical edge of the device or between active areas. To predict their impact on the breakdown voltage in correspondence of the isolation implants, the simulations proposed in Fig. 3.8 have been performed. The figure reports, in the *top* part, two cross-sections of the UFSD inter-pad region between active areas showing the field intensity map (in grey tones) simulated at the breakdown voltage. *Top left* panel refers to a structure with a very short metal overhang (with respect to the n^+-contact) deposited on top of the device, while the *top right* panel concerns the case of implementing a field plate. Furthermore, the *bottom* panels (c) and (d) show the $I(V)$ characteristics of the devices sketched, respectively, in subfigures (a) and (b). It's is evident that the use of a metal field plate fully covering the JTE implants is beneficial to keep under control the electric field, which relaxes in silicon. Indeed, the region with the highest field intensity moves from the junctions of the inter-pad implants to the oxide, in correspondence with the metallization edge, allowing to reach the breakdown at a voltage approximately 200 V higher.

Similar reasoning can be made if we want to characterize the field in the detector periphery, for example, in order to design more robust protection structures which

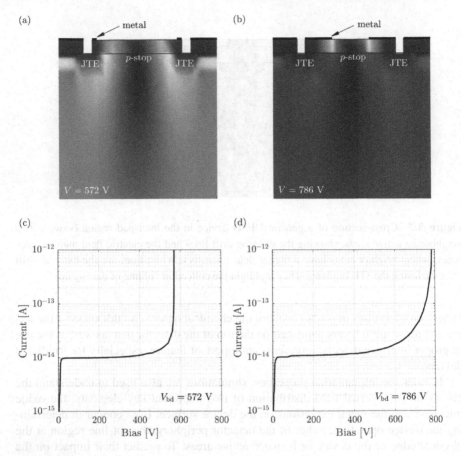

Figure 3.8 *Top* : simulated electric field intensity map (in grey tones - brighter areas have a higher intensity) at the breakdown voltage V_{bd}, for two UFSD with different inter-pad regions, short (a) or long (b) metal overhang (black pattern on the device surface). *Bottom*: corresponding $I(V)$ curves.

allow higher breakdown voltages. Overall, TCAD simulation offers a powerful tool to have relatively fast and reliable feedback on the physics driving each part of the system.

3.5.2 TRANSIENT PROCESSES

This section deals with one of the most interesting and crucial targets of simulating UFSDs: the signal formation. The UFSD operation mechanism is based on the multiplication by a certain gain factor (which is bias-dependent) of the primary *e-h* pairs produced when a charged particle crosses the detector.

The numerical implementation consists of a time-dependent process during which a heavy ion or a laser beam simulates a particle crossing the detector, with a custom trajectory and a well-defined energy released in the silicon lattice. Since this procedure, in principle, holds for each particle in any kind of semiconductor, it is important to characterize the detection process for a minimum ionizing particle (MIP) in silicon. A possible approach can be represented by the calibration of the collected charges in a detector without the gain implant. Here, the absence of multiplication allows to properly tune the energy released by the ion or the laser, allowing to accurately describe the gain in UFSD at low applied voltages, i.e., below the avalanche onset. Moreover, as explained in Section 2.10, after high values of fluence, the electric field in a PIN can be raised so much that charge multiplication happens in the sensor bulk. This aspect is important in the simulation of heavily irradiated UFSD.

Figure 3.9 shows two significant examples coming from the calibration campaign carried out on two independent productions of PIN devices, one by CNM and the second by HPK [36]. The number of charges has been obtained by integrating the signal response in time, both in simulations and in the laboratory measurements. In order to accurately fit the experimental data, the curves presented in the *left* panel have been simulated by setting a MIP releasing an energy equivalent to the production of ~ 60 electron-hole pairs per crossed micron whereas in the *right* panel we set ~ 70 pairs. Both devices are irradiated (respectively, with neutrons at a fluence of $3 \cdot 10^{15}$ n_{eq}/cm^2 and with pions at $1.5 \cdot 10^{15}$ n_{eq}/cm^2) so these simulations also accounted for the phenomenological model written in Eq. (3.76) for the acceptor creation/deactivation with fluence. The plots indicate that both Massey and van Overstraeten-de Man models are adequate in reproducing the collected charge while the Okuto-Crowell predicts a higher bias value for the onset of the multiplication in the bulk.

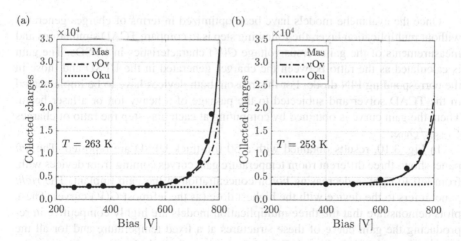

Figure 3.9 Measured (symbols) and simulated (lines) number of collected charges in 50 μm-thick PIN diodes manufactured by (a) CNM and (b) HPK. Simulations have been performed using three different avalanche models [52, 54].

It is worth stressing that, even with different formalism, the same result is obtained simulating either the injection of heavy ions or that of an infrared laser beam (for the wavelength 1064 nm) releasing the energy of 310 W/cm^2. The main difference between the two numerical approaches is that the simulation with ions is more suitable for a comparison with results obtained with particle while the simulation with laser, allowing to define the illumination over a window with finite width, is more appropriate with results obtained with the Transient Current Technique, TCT (see Section 4.3 and Ref. [74]).

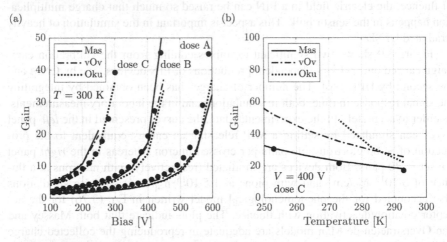

Figure 3.10 Measured and simulated gain value versus (a) bias and (b) temperature, in 50 μm-thick UFSDs with different gain implant dose (increasing from A to C) [52, 117].

Once the avalanche models have been optimized in terms of charges generated without multiplication layer, the following step is to compare TCAD simulations and measurements of the gain versus voltage $G(V)$ characteristics in UFSDs. The gain is calculated as the ratio between the charges generated in the UFSD and those in the corresponding PIN diode. For this reason, both devices have to be implemented in the TCAD solver and subjected to the passage of a heavy ion or a laser beam. Then, the gain curve is obtained by computing at each bias step the ratio of charges Q_{UFSD}/Q_{PIN}.

In Fig. 3.10, results obtained with a 50 μm-thick UFSD are reported. The *left* panel shows three different room temperature gain curves coming from devices with, from left to right, a decreasing boron concentration in the gain implant. The *right* panel refers to the device with the highest dose (as the leftmost $G(V)$ curve). These plots demonstrate that the three multiplication models are highly competitive in reproducing the gain curve of these structures at a fixed temperature and for all the gain implant doses considered. For what concerns the gain increase with temperature, the Massey model is the most reliable. The slope of $G(T)$ is quite satisfactory also for the van Overstraeten-de Man avalanche, while the Okuto model predicts a much steeper dependence.

The *left* panel of Fig. 3.11 presents a second comparison between simulated and measured $G(V)$ curves at room temperature, using sensors from the UFSD2 production. The gain implant has been obtained with boron or gallium at different doses, with/without the co-implantation of carbon atoms. For simplicity, only results obtained with the van Overstraeten-de Man model have been reported. The plot on the *right* panel reports the amplitude seen by two adjacent pads during a TCT position scan (see Section 4.3) obtained – both experimentally and numerically – by moving the spot of an infrared laser beam along the detector surface in correspondence of the inter-pad region.

(a)   (b)

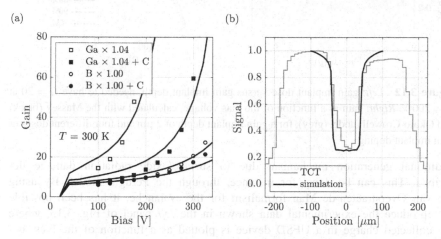

Figure 3.11 *Left*: measured and simulated (using the van Overstraeten-de Man avalanche) room temperature gain curves in 50 μm-thick UFSD differing in the gain implant dopants. *Right*: laser scan of the inter-pad region showing the normalized signal intensity versus the position of the laser spot on the detector surface [51].

The next step in the UFSD simulation is the calibration of the parameters characterizing the gain implant. Figure 3.12 shows the simulated effects of varying the gain implant dose and depth. The *left* panel reports how the gain boron dose (normalized at 1 μm) has to be scaled when implanting at different depths in order to have gain $G = 20$ at the reverse bias $V = 200$ V. As explained in Section 2.1.1, when the gain implant is deeper, the doping has to be decreased. The *right* panel of Fig. 3.12 shows how the gain changes varying the doping density of the gain implant. For a fixed depth of the boron profile, 2 μm, the implant dose has been varied from 100% to 85%. As the acceptor density becomes lower, the gain decreases. These predictions quantitatively depend on the avalanche model: in this example, the Massey (black curves) model gives higher gain with respect to the Okuto-Crowell model (grey curves).

One important ingredient of the UFSD design is simulating the effects of radiation on the detector performances. To this aim, it is necessary to account for the doping creation and deactivation, both in the gain implant and in the bulk, and the

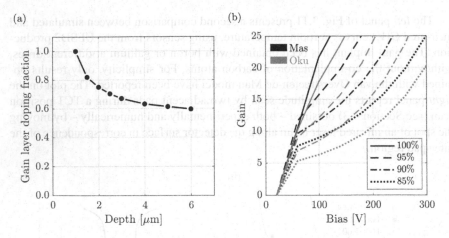

Figure 3.12 *Left*: gain implant dose versus gain implant depth, needed to have $G = 20$ at $V = 200$ V. *Right*: gain as a function of the bias voltage, calculated with the Massey (black) and Okuto-Crowell model (grey), for a gain implant depth of 2 μm and four different doses of gain implant doping.

additional generation mechanism due to surface/oxide radiation damage described. This can be done, for instance, through the Perugia model. By using the van Overstraeten-de Man formalism for the avalanche, it has been possible to reproduce the experimental data shown in the *left* panel of Fig. 3.13, where the collected charge in a UFSD device is plotted as a function of the bias, before irradiation and at two fluences. The numerical prediction is quite satisfactory, also in consideration of the changes induced by a temperature variation

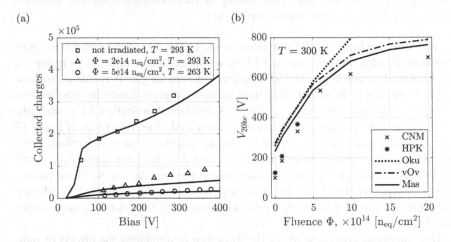

Figure 3.13 *Left*: number of collected charges versus bias for a 300 μm-thick UFSD before and after irradiation. *Right*: bias values needed to collect $2 \cdot 10^4$ electrons as a function of the fluence in 50 μm-thick UFSDs from different vendors [52].

(293 K vs 263 K), confirming the robustness of the TCAD approach in terms of radiation-hardness. Finally, in the *right* panel of Fig. 3.13, the measured and simulated bias to collect $2 \cdot 10^4$ electrons versus fluence is reported for several 50 μm-thick UFSDs. Here all the three avalanche models have been tested, obtaining that – as in the simulation of gain as a function of temperature – the Okuto–Crowell model is the worst choice to reproduce this set of measurements.

3.6 RESISTIVE AC-COUPLED SILICON DETECTORS DESIGN

As explained above, the UFSD design is optimized to achieve the best temporal resolution. Consequently, the signals generated by particles hitting in the inter-pad region should not be amplified to avoid out-of-time signals (see Section 2.8). The no-gain region decreases the detector fill factor and leads to the use of multiple staggered layers to achieve hermetic coverage. A possible strategy to overcome this issue is the AC-LGAD [61] paradigm, which consists in the implementation of a continuous gain implant that achieves 100% fill factor, as shown in Fig. 3.14. In order to

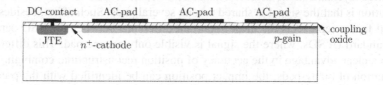

Figure 3.14 Schematic cross-section (not to scale) of an RSD (AC-LGAD) sensor.

reconstruct the hit position without segmentation, AC-LGADs make use of two key design elements: (*i*) a coupling oxide and (*ii*) a resistive n^+-cathode. For the signal to be visible on the AC-pads, both elements need to be correctly engineered so that the lowest impedance path to ground for the signal is via the read-out electronics. The signal discharges with an *RC* time constant given by the product of the AC-pad capacitance and the n^+-cathode resistivity. Thus, since the target is to have a discharge time long enough for the signal to be seen by the pads, the *RC* time constant must be chosen to be longer than the signal formation time ($\sim 1 - 2$ ns). However, to avoid pile-up effects, the *RC* should be sufficiently short to allow a prompt return to the baseline. The most important parameters are the oxide composition and thickness, determining the coupling strength, and the n^+ implant dose and profile, which instead directly modify the resistivity of the cathode. The geometrical configuration of the AC-pads (e.g., dimension and pitch) is also decisive since their geometrical dimension directly impacts the *RC* time constant. Given the large number of parameters concurring in the AC-LGADs design, it is crucial to have a reliable numerical tool for the prediction of signal formation and read-out properties.

In the last few years, several research groups began to develop and also fabricate AC-LGADs. In this chapter, a specific design, produced by INFN in Torino (Italy) and FBK, called Resistive AC-Coupled Silicon Detectors (RSD), is explained. The name, RSD, refers to its two main features: the resistive implant and the coupling oxide. Here, some parametric simulations referring to standard values are shown in order to demonstrate how TCAD can help in the design of such devices. To this aim, the impact of varying the RSD design parameters on the signal waveform is analyzed. Several simulated signals produced by a MIP in an RSD are reported in Fig. 3.15. In these results, the detector is composed of a row of three AC-pads, and the particle is always crossing the device in the center of the first pad and black, grey, and light grey lines refer, respectively, to the signal seen in the hit pad, its first and second neighbor pad. The preliminary observation coming from the simulations is that the current as a function of time has a bipolar behaviour, an intrinsic feature of the AC-coupled read-out paradigm.

The first lobe is generated by the coupling with the AC-pads, when the signal is collected by the resistive n^+-cathode, while the opposite lobe is due to their subsequent discharge to ground, which takes place through the DC-contact (see Fig. 3.14). The discharge characteristics – such as the amplitude and duration – depend on the RC constant of the equivalent read-out circuit, so, on the coupling capacitance of the oxide layer and on the sheet resistance of the cathode. The second fundamental observation is that the signal is shared among several pads. Such effect, besides the 100% fill factor, represents the most important difference between the RSD paradigm and standard UFSDs, where the signal is visible only on one pad. This difference leads to a clear advantage in the accuracy of position reconstruction: combining the information of many pads, the impact position can be identified with the precision of a few microns [55, 58]. As for the other figures of merit related to the signals, also the charge sharing depends on the properties of both the coupling oxide and the resistive implant.

The *top left* and *top right* panels of Figure 3.15 report the signals simulated with a 2D TCAD implementation, injecting 1 MIP in a 50 μm-pitch RSD device with 45 μm pad size. First of all, consider only the pad hit directly by the particle (black curves). The solid lines show the signals obtained implementing the standard values of the FBK technological parameters (low oxide thickness and low n^+-cathode dose). The dotted lines, instead, refer to an increase of either the oxide thickness (*top left*) or cathode resistivity (*top right*). The simulations show that both the peak amplitude and the discharge duration increase when the oxide is thinner (higher capacitance) or the n^+-cathode dose decreases (higher resistivity). The signals seen on the first and second neighboring pad, respectively in grey and light grey, have the same behaviour. The simulation, therefore, indicates that if the RC time constant is too short, the signal discharges before being fully formed.

The *bottom left* panel of Fig. 3.15 shows a different scenario. In this plot, the same RSD geometry of the previous simulations is compared to a modified version, where the pitch is doubled and the pad size is 95 μm. Moreover, the fabrication technology has been slightly changed, a 50% thicker oxide has been used. Focusing again on the signals coming from the pad crossed by the particle (black lines), it is evident that the bigger pad has a larger signal. Considering the adjacent pads, the first and second neighboring pads in the 100 μm-pitch RSD produce a signal which is lower in amplitude and broadened in time with respect to the 50 μm-pitch case.

(a) (b)

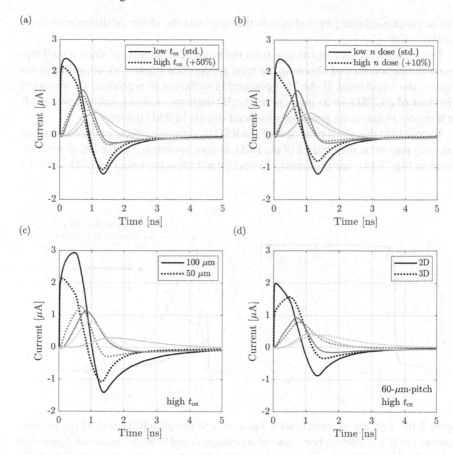

(c) (d)

Figure 3.15 Simulated signals in a three-pads RSD detector, biased at 300 V, at room temperature. Black lines represent the signal coming from the hit pad, while the different tones of grey refer to the first and second neighboring pad. *Top left*: signals in two RSDs differing for the thickness of the coupling oxide. *Top right*: signals in two RSDs differing for the n^+-cathode resistivity. *Bottom left*: signals in two RSDs differing for the sensor pitch. *Bottom right*: signals generated using either a 2D or 3D simulation [54].

Another important result stressing the importance of numerical simulations is presented in the *bottom right* panel of Fig. 3.15. In this plot, the 2D and 3D implementations of a 60 μm-pitch RSD are compared. Here the resistive n^+-cathode dose is the standard one, while the oxide has been chosen in its thick version. In these simulations, all the pads have in common the same behaviour: the signals computed with the 3D structure are slightly lower and longer with respect to the 2D case. Even if in the 2D condition the tool emulates the 3D scenario projecting the third dimension by a customizable factor, the full-3D simulation takes into account additional built-in volume-related physical effects that originate the observed differences in the signals.

This example is very instructive from the numerical standpoint since it well represents the importance of choosing the most appropriate framework according to the target of the simulation. If the 2D geometry is sufficient to reproduce the electrical behaviour of a UFSD, or its gain curve, the 3D implementation is necessary to study the transient phenomena generating induced signals in RSD detectors.

To conclude this short overview of the RSD design, the effects of the termination structures present at the borders of the RSD, in correspondence of the DC-contact (as shown in Fig. 3.14), are presented. To understand the edge effect in RSD, consider

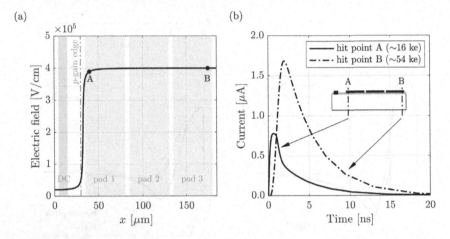

Figure 3.16 *Left*: electric field in a 1×3 array of a 50 μm-pitch RSD, with 45 μm pad size, biased at 160 V. The MIP has been injected in positions A and B. *Right*: simulated signals seen on the DC contact of the same device as a function of the distance from the sensor edge [54].

the simulation of a 1×3 array with 50 μm-pitch and 45 μm pad size through a 2D implementation again. A MIP is injected in the first pad, 5 μm from its left edge, and in the third one, 5 μm before its end. The electric field beneath the silicon/oxide interface as a function of position is reported in the *left* panel of Fig. 3.16. At small x values, where the DC-contact is located, the field is of the order of few tens of kV. Then, the field experiences a steep increase, starting at the edge of the multiplication implant, located at 30 μm from the origin (the greater the lateral spread of this implant, the earlier the field increase). Past this sharp rise, the field increases slowly: at the point A, at 40 μm, is 394 kV/cm while at the point B, at 175 μm, is 404 kV/cm. The *right* panel reports the simulated signals collected by the DC-contact at 160 V, when the MIP crosses the device in A and B. The inset drawing shows the cross-sectional view of the simulated RSD, as well as the two tracks crossing the detector perpendicularly to the surface. The signal from A arrives earlier than that from B, since the discharge path from the hit point to the contact is shorter, however, the number of collected charges is smaller (respectively, 1.6 against $5.4 \cdot 10^4$ electrons) since the multiplication field is lower in A than in B. This simulation highlights that termination structures may affect the detector response, creating an area of lower gain at the periphery of the active region. This is why numerical simulations are essential in designing the RSD and, in particular, in defining specific layout rules, such as the minimum distance between the peripheral pads and the gain implant edge.

4 Experimental Techniques

This chapter describes the experimental techniques most commonly used in the characterization of silicon sensors. The description focuses on the experimental setups and methods, which are referred to in many of the measurements reported in the following chapters.

4.1 STATIC CHARACTERIZATION OF UFSD SENSORS

The static characterization of a silicon sensor, with or without internal gain layer, is carried out by measuring the current-voltage ($I(V)$), capacitance-voltage ($C(V)$), and capacitance-frequency ($C(f)$) characteristics of the device in absence of external particles.

The typical setup for these measurements consists of a probe station, with or without the capability to perform measurements at controlled, and eventually low, temperature, electrically connected to a set of devices (power supply, CV-meter, etc.). Alternatively, these devices can be replaced by a single mainframe device analyzer equipped with specialized modules (high/low power, high/low current, capacitance, etc.)

The central part of the probe station, called chuck, is a support where the device under test (DUT) is placed and kept still by a vacuum system. An optical microscope, equipped with different magnification optics, and a video camera positioned in a dedicated opening, are necessary to visualize the DUT on the chuck. The chuck usually provides the bias voltage to the DUT. Tungsten-rhenium needles are available for contacting specific points on the front side of the DUT; the positioning of the needles occurs with manipulators equipped with micro-metric screws. Chuck and needles are electrically connected with the measuring instruments. In the case of n-in-p devices, as the UFSDs described in this book are, the chuck is biased negatively, and the needles are at zero voltage.

In the following paragraphs, three devices are used in the description of the setups (generically called *SMU* as they can source and monitor electrical quantities):

1. High Voltage Source Monitor Unit (HV-SMU), able to force voltage or current and simultaneously measure voltage and/or current, in a range up to ~ 1–2000 V/5–10 mA;
2. Medium Power Source Monitor Unit (MP-SMU), as above but working in a range of voltage and current up to ~ 100 V/0.1 A, and with a minimum measurement resolution of 10–100 fA/0.5–1 μV;
3. Multi-Frequency Capacitance Measurement Unit (MF-CMU), able to issue different signal amplitudes and different frequencies, with a frequency range 1 kHz–5 MHz and an AC-signal level up to 250 mV;

Current-Voltage setup

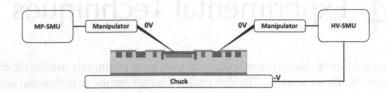

Capacitance-Voltage/Capacitance-Frequency setup

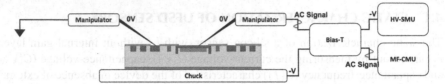

Figure 4.1 Schematic representation of a probe station configured for a current-voltage measurement (*top*) and for a capacitance-voltage/capacitance-frequency measurement (*bottom*).

4.1.1 CURRENT-VOLTAGE MEASUREMENT

When a reverse bias is applied to a *pn* junction, a steady current, called *leakage* or *dark* current, is flowing through the device, even in absence of an external stimulus. Surface, bulk, and avalanche currents contribute to the total leakage current.

The leakage current of a single pad device as a function of the reverse bias is measured using the $I(V)$ setup shown in Fig. 4.1 (*top*). The high voltage and ground leads of the HV-SMU are connected to the chuck and the guard-ring, respectively, while a second needle, attached to the MP-SMU, is contacting the pad under test. The MP-SMU has a higher current resolution than HV-SMU, therefore it is better suited to measure the leakage current flowing in the pad. In the case of multi-pad devices, all the pads should be contacted with needles. The pads under test are wired to the available MP-SMUs, while the remaining pads are short-circuited to the guard-ring. In this configuration, it is possible to measure the $I(V)$ characteristics of several pads in a single bias voltage scan.

The $I(V)$ curves of a PIN diode and of a UFSD are compared in Fig. 4.2. The $I(V)$ characteristic (dashed line) of a PIN diode in reverse polarization follows the behaviour predicted by the Shockley diode equation [129]: the leakage current reaches a plateau value, temperature-dependent, and it remains roughly constant up to the breakdown voltage (not visible in the plot). The UFSD $I(V)$ characteristic (solid line), instead, has an exponential trend above a certain bias voltage value due to the presence of the gain layer. At low voltages, the $I(V)$ curve differs from that of a PIN for the presence of a knee, at about a few tens of volts, representing the depletion voltage of the gain layer.

The $I(V)$ curves are a powerful tool for sensor testing and characterization. They carry very useful information on the sensor type, and almost every manufacturing

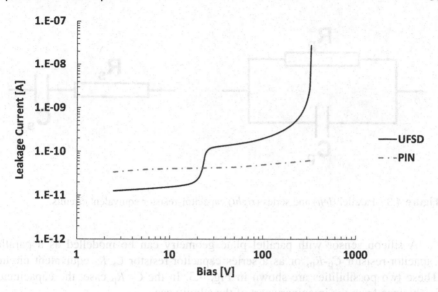

Figure 4.2 Current-voltage characteristic curves of a 50 μm-thick UFSD (solid line) and PIN diode (dashed line).

problem leads to a deviation from the expected shape. For UFSD sensors, the $I(V)$ curve allows for identifying the gain layer characteristics. The gain layer depletion voltage is related to the gain layer profile, and the exponential growth of the leakage current is gain dependent.

4.1.2 CAPACITANCE-VOLTAGE MEASUREMENT

The study of the DUT capacitance as a function of the bias voltage $C(V)$ and of the frequency $C(f)$ finds extensive application in the qualification of semiconductor and multiple-layered structures. Figure 4.1 (*bottom*) shows, for a single pad device, the setup for these two types of measurements using the HV-SMU and the MF-CMU. The two modules are interfaced via a bias-T, which merges the DC bias voltage, provided by the HV-SMU, and the AC-signal, from the MF-CMU. The bias-T outputs one high ($\pm V$) and one low (0 V) DC voltage level, with the AC-signal superimposed. The chuck and the needle contacting the pad are wired to the high and low voltage levels, respectively, while the guard-ring is grounded. In this example, the value of the capacitance between the pad and the backplane is measured. In a multi-pad device, the measurement of the total capacitance of a pad, i.e., the sum of the capacitance to the backplane and all its neighbors, requires some ingenuity. Usually, this measurement is done in two (or more) steps, measuring a given part of the total capacitance in each step. The contribution due to the backplane is measured with the setup just described; the capacitance between pair of pads is measured by connecting the two leads of the MF-SMU to the two pads under study, biasing the sensor with the HV-SMU. Care should be taken to appropriately refer the HV-SMU and MF-SMU to a common ground value.

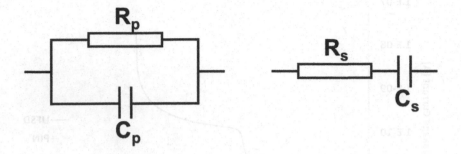

Figure 4.3   Parallel (*left*) and series (*right*) capacitor-resistor equivalent circuits.

A silicon sensor with parallel plate geometry can be modelled as a parallel capacitor-resistor C_p-R_p, or as a series capacitor-resistor C_s-R_s equivalent circuit. These two possibilities are shown in Fig. 4.3. In the C_p-R_p case, the capacitance is obtained from the imaginary part of the admittance

$$Y = \frac{1}{R_p} + j\omega C_p, \tag{4.1}$$

dividing it by $\omega = 2\pi f$, where f is the frequency of the probe AC-signal. When using the C_s-R_s model, the capacitance is obtained from the imaginary part of the impedance:

$$Z = R_s - j\frac{1}{\omega C_s} = \frac{1}{Y}. \tag{4.2}$$

The C_p-R_p and C_s-R_s equivalent circuits are good models for unirradiated sensors: the leakage current of these devices is very low, implying that the conductivity is low and that the measured admittance Y or impedance Z are given almost exclusively by the capacitance of the device. In irradiated sensors, the high leakage current implies that the C_s-R_s model cannot describe the correct behaviour of the device since it does not allow for the presence of leakage current.

Figure 4.4 shows the $C(V)$ curves of a PIN diode and of a UFSD device. The UFSD curve (solid lines) shows a sudden capacitance drop (knee) at a few dozen volts, corresponding to the depletion voltage of the gain layer region, V_{GL}. V_{GL} is proportional to the active doping concentration N_A, to the square of the gain implant width w, and to the gain implant depth d:

$$V_{GL} \propto (1 + 2\frac{d}{w})N_A w^2. \tag{4.3}$$

Additional considerations on the meaning of V_{GL} are provided in Section 5.1.1.

The full depletion voltage V_{FD} of the device occurs when the capacitance becomes constant. The difference between V_{FD} and V_{GL} is the depletion voltage of the

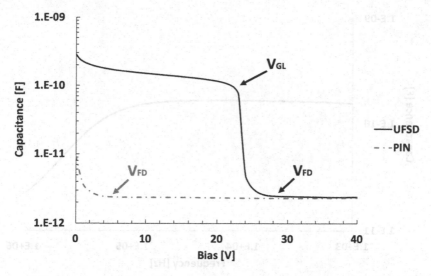

Figure 4.4 Capacitance-voltage characteristic curves of a 50 μm-thick UFSD (solid line) and PIN diode (dashed line).

remaining part of the bulk, called V_{bulk}. In case of very thin gain layer region (a few microns), V_{bulk} is very similar to the full depletion voltage of the PIN diode (dashed line).

The $C(V)$ measurements use a probe AC-signal at a fixed frequency. The identification of the optimal AC-signal frequency is achieved by performing a capacitance scan as a function of the frequency, with the setup shown in Fig. 4.1 (*bottom*). The sensor can be approximated to an RC network, with a frequency-dependent behaviour similar to that of a low pass filter. The $C(f)$ measurement in Fig. 4.5 shows, for the UFSD under test, a capacitance more or less constant up to a frequency of ~ 100 kHz, while it decreases above this value. The decrease of the measured capacitance is caused by the shunting of the test frequency to ground before the full sensor volume can be explored. The optimal test frequency is therefore to be selected where the capacitance measurement is constant.

It is important to note that the optimal AC-signal frequency value might change by changing the DUT or even for the same sensor but under varied measurement conditions. Figure 4.6 (*top*) shows, for instance, the $C(f)$ curves of two devices with depleted regions of different resistivity (gain layer of an LGAD and bulk of a PIN): the capacitance of the bulk (high resistivity) is independent of the test frequency in the 1 kHz–1 MHz range, while the capacitance of the gain layer (low resistivity) is frequency-dependent. In addition, the $C(f)$ characteristic of irradiated devices differs from that of unirradiated devices, as it is affected by the number of defects/traps in the silicon lattice. Figure 4.6 (*bottom*) shows the $C(f)$ curves of five 50 μm-thick UFSD sensors, one unirradiated and 4 irradiated, the highest at a fluence of $3 \cdot 10^{15}$ n_{eq}/cm^2. They are measured at room temperature and in a condition of partial depletion voltage of the gain layer ($V = -10$ V). The plot shows that the frequency range suitable

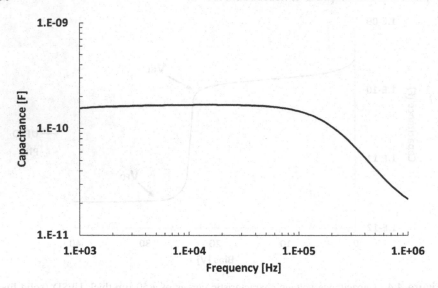

Figure 4.5 Capacitance-frequency characteristic of a 50 μm-thick UFSD at a fixed reverse bias voltage of −10 V.

to perform $C(V)$ measurements is much reduced for the irradiated samples. Other studies too confirm that it is advisable to select lower frequencies and temperatures when testing irradiated devices [59]. The $C(f)$ characterization is, therefore, an essential step to correctly perform a $C(V)$ measurement.

An alternative method to obtain a $C(V)$ characteristics is to use the so called Quasi-Static Capacitance Voltage (QS-CV) method. As compared to a high-frequency $C(V)$ measurement, its most important benefit is the capability of measuring slow interface states at the semiconductor-dielectric interface. In the QS-CV step voltage method, the bias is increased in steps and the capacitance is obtained by measuring the collected charge: $C = \Delta Q / \Delta V$. Given the presence of the sensor leakage current, a correction mechanism for the value of Q needs to be implemented. This correction can be done, for example, measuring the leakage current for a given time interval before and after each step.

Using the information provided by the $C(V)$ measurements, it is possible to:

1. extract the gain layer implant profile (amplitude, width, depth) by computing the active doping concentration $N_A(d)$ as a function of depth d. This is obtained by first calculating d from the value of C with

$$d = \frac{\varepsilon_{Si} A}{C} \tag{4.4}$$

and then the doping density at that depth d with:

$$N_A = \frac{2}{\varepsilon_{Si} q A^2 \frac{\partial 1/C^2}{\partial V}}, \tag{4.5}$$

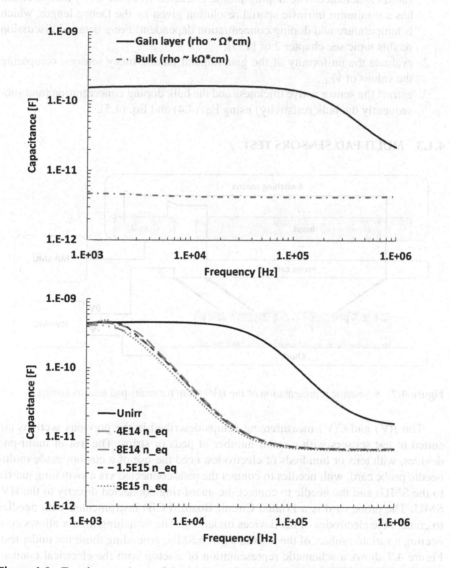

Figure 4.6 *Top*: the capacitance-frequency curves of 50 μm-thick UFSD (solid line) and PIN diode (dashed line) biased at −10 V. *Bottom*: room temperature capacitance-frequency curves of UFSD sensors irradiated with neutrons up to a fluence of $3 \cdot 10^{15}$ n_{eq}/cm^2, biased at −10 V.

where ε_{Si} is the silicon dielectric constant, q is the elementary charge, A is the active area of the sensor, d is the width of depleted region and C is the measured capacitance. The doping profile extracted from the $C(V)$ measurement has a minimum intrinsic spatial resolution given by the Debye length, which is temperature and doping concentration dependent. For a detailed discussion on this topic see chapter 2 of [129];

2. evaluate the uniformity of the gain implant among many sensors, comparing the values of V_{GL};

3. extract the sensor active thickness and the bulk doping concentration (and subsequently the bulk resistivity) using Eq. (4.4) and Eq. (4.5).

4.1.3 MULTI-PAD SENSORS TEST

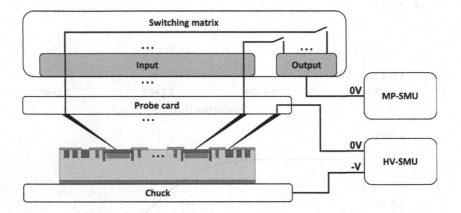

Figure 4.7   Schematic representation of the $I(V)$ setup for multi-pad sensors testing.

The $I(V)$ and $C(V)$ measurement setups described in the previous sections are suited to test sensors with a small number of pads or strips. The test of multi-pad devices, with tens or hundreds of electrodes, need the use of a custom made multi-needle probe card, with needles to contact the pads connected via a switching matrix to the SMUs and the needle to contact the guard-ring connected directly to the HV-SMU. The probe card is a Printed Circuit Board (PCB) instrumented with needles to contact the electrodes of the devices under test. The switching matrix allows connecting a variable subset of the pads to a given SMU, grounding those not under test. Figure 4.7 shows a schematic representation of a setup with the electrical connections needed for $I(V)$ measurements of a multi-pad device.

4.2 CCD-CAMERA SETUP

The CCD-camera used in this setup employs a cooled, ultra low-noise Charge-Coupled Device (CCD) to detect very faint photons emission in the visible spectrum. When imaging silicon sensors, such a camera allows to identify approximately the position where the breakdown occurs on the sensor. Near breakdown conditions, high electric fields are present on the surface, between the termination structures. These high electric fields generate surface charge carriers, whose recombination leads to photons emission (called in this context 'hot spots'). In this setup, the CCD-camera is positioned in the camera opening of the probe station microscope, and the system is placed inside a dark box.

Figure 4.8 shows two typical images acquired with a CCD-Camera setup: the hot spots are clearly visible since, in the dark, they generate more photons than the rest of the sensor. For this reason, they appear as white spots. In both images, the hot spots are located between guard-ring structures on the surface of the sensors under test.

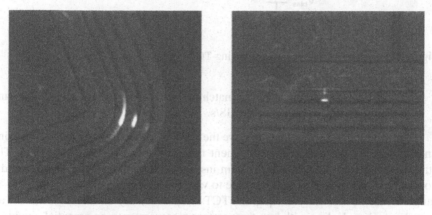

Figure 4.8 CCD images of two different portions of a UFSD sensor. In both locations, the hot spots are visible between guard-ring pairs.

4.3 TRANSIENT CURRENT TECHNIQUE SYSTEM

The Transient Current Technique (TCT) has been extensively used to characterize silicon sensors in the past two decades. In this section, the principles of operation of this technique are presented, along with the descriptions of some typical measurements that can be performed on UFSD sensors.

The transient current technique exploits the signal induced by the motion of free carriers in a semiconductor (see Section 1.2): focused laser pulses generate e-h pairs, which induce a current signal on the read-out electrode. The signal is amplified and stored in an oscilloscope or digitizer for offline analysis (Fig. 4.9). Since 50 μm-thick UFSDs produce very short signals, about 1 ns, the TCT setup needs to be optimized with:

1. laser shot with a duration of ~ 50–100 ps and selectable intensity to simulate the signal generated by a MIP traversing the sensor;
2. a 20–40 dB current-mode amplifiers with a high bandwidth (about 1–2 GHz);

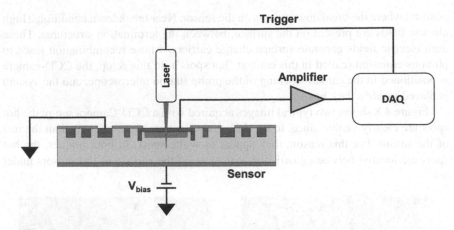

Figure 4.9   Schematic diagram of a scanning-TCT system in Top-TCT mode.

3. oscilloscope or digitizer with a matching bandwidth (at least a few GHz) and a sampling rate in excees of 20 GS/s.

The scanning-TCT system can move the DUT with micrometric precision, using an x-y stage with a maximum movement range of about 10–20 centimeters. The stage is equipped with a cooling system instrumented with a Peltier element and a cooled block connected to a chiller able to vary the temperature in the -20–$+80\,°C$ range. It may be necessary to equip the TCT setup with a dry-air inlet to lower the air dew point when dealing with low-temperature measurements, an essential condition for irradiated sensors testing. The TCT setup has an optical laser focusing system, mounted perpendicularly to the x-y stage (in the Top-TCT mode), on a z-translator with micrometric precision of movement. This system allows achieving a laser beam spot of about 10 microns inside the sensor [108, 121].

The laser wavelengths suitable for the characterization of UFSD sensors are in the 400–1060 nm range. The infrared (IR) laser ($\lambda = 1060$ nm), with an absorption length in silicon of ~ 1 mm, crosses with little attenuation the DUT. The blue laser ($\lambda = 400$ nm), with an absorption length in silicon of few microns, simulates the energy deposition of α-particles.

There are several differences in the creation of e-h pairs in the sensor bulk by a laser beam or by a MIP particle: (i) the total number of e-h pairs generated by a laser beam is almost constant, while that of a MIP follows the Landau distribution, (ii) the density of e-h pairs is constant along the laser track while along the MIP track is non-uniform, and (iii) the laser beam creates e-h pairs within a cylinder of about 10 μm diameter (depending on the laser optics), while for a MIP this diameter is much smaller, resulting in more substantial screening effects.

Figure 4.10 shows a typical signal generated by IR laser ($\lambda = 1060$ nm) in a 50 μm-thick UFSD, using a Top-TCT setup. As the laser intensity can be unstable during measurement, it is necessary to monitor the laser fluctuations using an appropriate control system. This system consists of an optical splitter and a reference

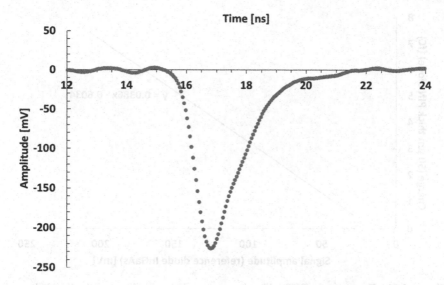

Figure 4.10 Signal generated in a 50 μm-thick UFSD using a Top-TCT setup with 1060 nm pulsed laser. The signal has been amplified by a current-mode amplifier with gain 40 dB and sampled by a 20 GS/s oscilloscope.

diode, sensitive to the laser wavelength in use. For example, the intensity of an IR laser can be split 10–90%, sending 10% of the signal to the reference diode. In the presence of laser intensity fluctuations, it is necessary to correct the collected charge Q by a factor, which can be obtained from a calibration curve, as shown in Fig. 4.11. The plot shows the relationship between the collected charge in a 50 μm-thick PIN diode and the amplitude of the signal induced in an InGaAs reference diode when using an IR laser.

Several key physical quantities can be extracted from the analysis of UFSD pulse shapes obtained on a TCT setup.

1. **Charge Collection Efficiency**: this measurement is performed by measuring the collected charge as a function of a given sensor condition (irradiation, bias, temperature, etc.). How to translate the signal seen on an oscilloscope into a charge value depends on the type of electronics used. If the amplifier works in current mode, then the charge Q can be obtained as:

$$Q = \frac{A_{\text{signal}}}{G_{\text{A}} \cdot R_{\text{in}}}, \tag{4.6}$$

   where A_{signal} is the area of the induced signal, G_{A} is the gain amplification chain, and R_{in} is the input impedance of the acquisition instrument (oscilloscope, digitizer, etc.). If, instead, the amplifier works in voltage-mode, then the charge is proportional to the signal amplitude, where the proportionality constant includes the sensor capacitance. In this case, the signal peak, not the area, is proportional to the signal charge.

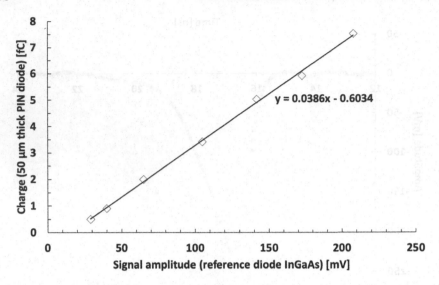

Figure 4.11 Example of a TCT calibration curve: relationship between the signal charge in a 50 μm-thick PIN diode and the signal amplitude in the reference diode (InGaAs) when using and IR laser.

2. **Sensor Gain**: the gain G is defined as the ratio between the amount of charge collected in a UFSD (Q_{UFSD}) and the amount of charge collected in a PIN diode (Q_{PIN}) with the same active thickness, and in same measurement conditions: irradiation, temperature, amplification chain, laser intensity, pulse duration, and bias voltage:

$$G = \frac{Q_{UFSD}}{Q_{PIN}}. \tag{4.7}$$

An important setting in this measurement is the laser intensity. If the laser is set to provide the same signal amplitude of a MIP signal, then the charge density inside the bulk is much lower than that generated by a MIP (given the much larger volume of the laser signal). In this condition, possible charge screening effects are much smaller during the laser studies. On the other hand, if the laser signal is set to generate the same charge density of a MIP, the overall signal amplitude becomes very large, changing the working point of the electronic chain. Scanning several laser intensities allows to identify the systematic effects and to extrapolate what is the gain obtained with a MIP signal.

3. **Inter-pad width**: the no-gain distance between two adjacent pads or strips is obtained by performing a laser scan along a line crossing the inter-pad region, and acquiring the collected charge profiles of the two pads as a function of the laser position. It is advisable to acquire many scans, at least 100–200, per each inter-pad measurement. For this specific measurement, the laser spot should be as small as possible (~ 10 μm or less) since its geometrical extension

interferes with the measurement precision. The collected charge in each of the two pads as a function of the laser position has the shape of an s-curve. Under the assumption that the gain region stops abruptly at the end of the gain implant (that is not entirely correct as shown in Fig. 3.7), the measured s shape curve can be understood as the convolution of a step function, representing the gain region, with a Gaussian function, representing the intensity profile of the laser pulse (see Fig. 4.12). In this convolution, the edge of the gain implant is the value of the x-axis at the 50% point amplitude, and the no-gain area is obtained as the difference between the two 50% points.

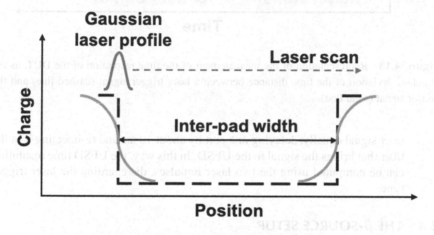

Figure 4.12 Sketch of the inter-pad width measurement, based on a laser scan between two adjacent pads or strips. The solid lines represent the profiles of collected charge in two adjacent electrodes, while the dashed line represents the physical no-gain inter-pad region.

4. **Time resolution - jitter**: the jitter component of the UFSD time resolution is usually measured as the standard deviation σ_t of the difference in time between the induced signal in the sensor and the trigger pulse generated by the laser controller, as shown in Fig. 4.13. The term σ_t can be written as

$$\sigma_t = \sqrt{\sigma_{t_{\mathrm{UFSD-jitter}}}^2 + \sigma_{t_{\mathrm{trigger}}}^2}, \tag{4.8}$$

where $\sigma_{t_{\mathrm{trigger}}}$ and $\sigma_{t_{\mathrm{UFSD-jitter}}}$ are the uncertainties on the trigger and UFSD time measurements, respectively. σ_t can be considered a good approximation of the UFSD jitter time resolution under the condition $\sigma_{t_{\mathrm{trigger}}} \ll \sigma_{t_{\mathrm{UFSD}}}$. For this measurement, it is important to configure the IR laser intensity to generate the same amount of charge induced by one MIP (for example the MPV) since the resolution is function of the signal amplitude. The time resolution measured using a laser pulse measures the jitter term of the total time resolution of the device (Eq. (2.12)) since the charge deposited by a laser pulse is not affected by Landau fluctuations. If the laser triggering system is affected by large uncertainties, the UFSD time resolution can be obtained by optically splitting the

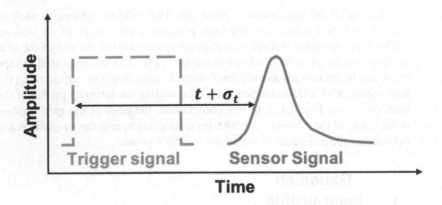

Figure 4.13 Representation of the measurement of the time resolution of the DUT, as the standard deviation of the time distance between a laser trigger signal (dashed line) and the sensor signal (solid line).

laser signal equally, delaying one part by about 10 ns, and re-injecting it in the fiber that brings the signal to the UFSD. In this way, the UFSD time resolution can be computed using the two laser impulses, disregarding the laser trigger time.

4.4 THE β-SOURCE SETUP

The experimental setup which uses β-particles as source of test particles, also known as β-source setup, is extensively used to study the timing performance of silicon sensors.

The reference time is provided by a sensor with known timing performances, aligned in space with the UFSD under test and the β-source. This sensor should have a time resolution as good as, or better than, that of the DUT to not spoil the measurement. Figure 4.14 shows a schematic block diagram of a β-source setup where two sensors are installed and readout concurrently, one of which serves as a trigger. The β-particles used in these measurements must be MIP, otherwise, the measured resolution will not be indicative of the sensor performances. A β setup with three planes, the DUT, the time reference, and a trigger sensor, ensure that the β particles are MIP since they cross through both the DUT and the time reference without being absorbed. If a setup with only two planes is used, DUT and trigger sensor, then in the offline analysis is necessary to limit the range of signal amplitudes in the trigger plane to a well-defined interval, where the trigger time resolution is well known and as much as possible constant.

The element ^{90}Sr is often used in β setups. It emits electrons with an energy of 0.546 MeV and of 2.28 MeV: the less energetic β-particles don't reach the second plane, while the most energetic ones pass through the device, generating a signal also in the trigger sensor. The read-out of the DUT and trigger signals are typically

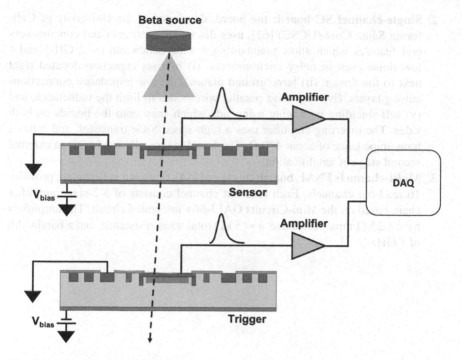

Figure 4.14   Schematic block diagram of the setup for time resolution measurements using β source.

performed using dedicated read-out boards with an amplification chain optimized for timing (see the following section). The β-setup can also be placed inside a cold box or climatic chamber to perform characterization at different temperatures.

4.5   UFSDS READ-OUT ELECTRONIC BOARDS

As explained in Section 2.6, the best timing performances are obtained when UFSDs are coupled to low-noise, low-impedance, fast amplifiers, working in current-mode. These characteristics are somewhat challenging to achieve and mostly not readily available in commercial amplifiers. For this reason, a few dedicated read-out boards have been designed with the intention of providing the best possible environment where to test the ultimate performances of the UFSDs. The following shortlist describes the read-out boards used to obtain most of the results presented in the following chapters.

1. **Bias-only board:** this type of board does not provide signal amplification but only filters for the bias voltage. The board normally has several connectors to link the read-out pads to external amplifiers. The bias-only board is typically used in experimental setups not dedicated to timing studies, for example, for inter-pad or gain studies with the TCT setup. For these measurements, a slower rising edge and a higher noise are not relevant.

2. **Single-channel SC board:** the board, developed at the University of California Santa Cruz (UCSC) [63], uses discrete components and contains several features which allow maintaining a wide bandwidth ($\sim$ 2 GHz) and a low noise even in noisy environments: (i) by-pass capacitors located right next to the sensor, (ii) large ground planes, (iii) low impedance connections among layers, (iv) very short parallel wire-bonds to limit the inductance, and (v) self-shielding packaging using lids which snap onto the boards on both sides. The inverting amplifier uses a high-speed SiGe transistor, and it has a trans-impedance of about 470 Ω. The board is typically used with an external second stage of amplification.

3. **Multi-channels FNAL board:** this board [14], designed at Fermilab, provides 16 read-out channels. Each read-out channel consists of a 2-stage amplifier chain based on the Mini-Circuits GALI-66+ integrated circuit. The amplifiers have a 25 Ω input resistance, a $\sim$ 5 kΩ total trans-resistance, and a bandwidth of 1 GHz.

5 Characterization of UFSDs

The properties of Ultra-Fast Silicon Detectors depend upon several basic parameters: the depth and width of the gain implant, the bulk active thickness and doping, the design of the inter-pad region, and the temperature of operation. The first part of this chapter analyses the impact of these quantities on the UFSDs performances. In the second part of the chapter, the temporal resolution of UFSDs is presented. The closing part of the chapter discusses the issues connected with the production of large quantities of UFSDs. The description of the UFSD productions mentioned in the following is reported in Appendix A.

5.1 GAIN LAYER CHARACTERIZATION

The core of a UFSD is its gain layer. This section analyses various aspects of the gain layer design, and how they impact the performances in term of temporal resolution, uniformity, and radiation hardness.

5.1.1 GAIN LAYER DESIGN

In the standard n-in-p UFSD design, the gain implant is obtained by the implantation and activation of acceptors. The implantation energy defines the depth of the implant, the activation temperature its width. Table 5.1 reports six gain layer designs with different combinations of the type of acceptor dopant, depth of implant, and thermal load of the activation.

Table 5.1
Gain Layer Characteristics in Different UFSD Productions

Type	Acceptor	Depth of Implant	Thermal Load
Shallow-BL	Boron	Shallow	Low
Shallow-BH	Boron	Shallow	High
Shallow-Ga	Gallium	Shallow	Low
Deep-BH	Boron	Deep	High
Deep-BL	Boron	Deep	Low
Broad-BH	Boron	Broad	High

The implant depth labels, *shallow* or *deep*, generically indicate implants around either 0.5–1 μm or 1.5–2 μm deep, while the label *broad* indicates an implant that starts at the *pn* junction and extends for a few microns. The thermal load label, low (L) or high (H), refers to the temperature used in the activation and diffusion of the acceptors. The $C(V)$ characteristics of these devices show that the gain layer depletion voltage, V_{GL}, varies in the range 20–60 V, Fig. 5.1. The gain implant profile of each design, shown in Fig. 5.2, has been extracted from the $C(V)$ characteristics, applying Eq. (4.4) and Eq. (4.5). The y-axis reports the doping density, while the x-axis the depleted region width. Numerically, the depleted region width is not exactly equal to the implant depth due to the undepleted n^{++} electrode. Both axes are in a linear scale. The peak doping densities are positioned at about 0.5–2.5 μm, with a doping concentration of $1-10 \cdot 10^{16}$ atoms/cm^3.

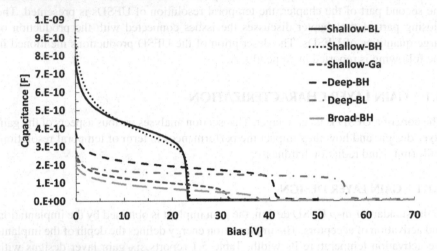

Figure 5.1 $C(V)$ measurements of six 50 μm-thick UFSDs with different gain layer designs: Shallow-BL, Shallow-BH, Shallow-Ga, Deep-BL, Deep-BH and Broad-BH.

A few considerations on Fig. 5.2:

1. The low-diffusion implants are narrower and higher than high-diffusion implants, both in the shallow and deep cases. The higher thermal load used in the high-diffusion case widens and lowers the boron implant profile.
2. The shallow-Ga implant is deeper than shallow-BL/H since gallium has been implanted at higher energy than boron.
3. The shallow-Ga implant is wider than then shallow-BL/H implants; gallium has a higher diffusivity than boron, so under the same thermal load a Ga implant widens much more than a B implant.
4. Deep-BL is deeper than deep-BH due to a higher implant energy.

The gain layer depletion value V_{GL} depends upon three parameters: the amount of dopant N_A, the implant width w, and its depth d, Fig. 5.3. The voltage V_{GI} needed to deplete the gain implant is given by:

$$V_{GI} = \frac{qN_Aw^2}{2\varepsilon}. \tag{5.1}$$

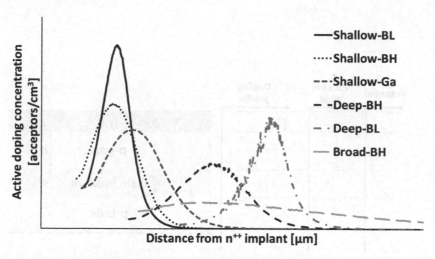

Figure 5.2 Profiles of various gain implants, extracted from the $C(V)$ measurements of Fig. 5.1, using Eq. (4.4) and Eq. (4.5).

This potential generates an electric field equal to:

$$\mathscr{E}_{\text{Gap}} = \frac{dV_{\text{Gap}}}{dw} = 2\frac{qN_A w}{2\varepsilon}. \tag{5.2}$$

The electric field $\mathscr{E}_{\text{Gap}}$ inside the gap can be considered constant since the doping in the gap d between the gain implant and the n^{++} electrode is many orders of magnitude lower than that in the gain implant. Therefore, the voltage drop over the gap d is given by:

$$V_{\text{Gap}} = \mathscr{E}_{\text{Gap}}d = 2\frac{V_{\text{GI}}}{w}d. \tag{5.3}$$

Summing the contributions of the gap and gain implant, the value of V_{GL} is obtained as:

$$V_{\text{GL}} = V_{\text{GI}} + V_{\text{Gap}} = V_{\text{GI}}(1 + 2\frac{d}{w}). \tag{5.4}$$

Figure 5.3 shows the trend of the electric field and of the potential inside the gain region while Table 5.2 compares the calculated and measured V_{GL} for the curves shown in Fig. 5.2. The V_{GI} and V_{Gap} values are calculated for a rectangular gain implant. The gain implant width, equal to the implant full-width half-maximum (FWHM), and the depth of implant, equal to the position of the maximum, are extracted from Fig. 5.2.

The comparison between V_{Gap} and V_{GI} indicates that the voltage needed to deplete the gain layer is mostly due to the electric field in the gap d, even for shallow gain implants. Equation (5.4) shows that, for equal implant geometry, V_{GL} increases linearly with twice the implant depth d: high V_{GL} ($\sim$ 40–50 V) are associated with deeper and not more doped implants.

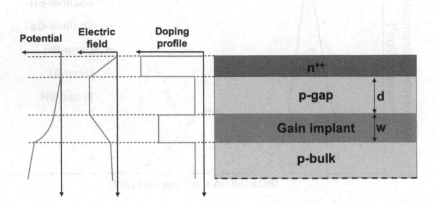

Figure 5.3 Electric potential and electric field corresponding to a given doping profile of the gain layer region.

Table 5.2

Calculated and Measured V_{GL} for the Curves Shown in Fig. 5.2. V_{GI} and V_{Gap} Are Calculated Assuming a Rectangular Gain Implant with Height, Depth and Width Taken from Fig. 5.1

Type	$V_{GI}^{Calc.}$ [V]	$V_{Gap}^{Calc.}$ [V]	$V_{GL}^{Calc.}$ [V]	$V_{GL}^{Meas.}$ [V]
Shallow-BL	5.7	17.1	22.8	22.4 ± 0.1
Shallow-BH	8.7	14.3	23	22.6 ± 0.1
Shallow-Ga	10.5	15.4	25.9	30.5 ± 0.1
Deep-BH	9.4	30.7	40.1	40.4 ± 0.1
Deep-BL	4.8	43.2	48	55.2 ± 0.1
Broad-BH	39.7	0	39.7	37.1 ± 0.1

5.1.2 IMPACT OF THE GAIN LAYER DESIGN ON PERFORMANCES

In the following section, a few of the parameters of the gain layer design are analyzed: narrow versus wide gain implants, deep versus shallow gain implants, high versus low gain implant doping densities, and the effect of a thin versus thick sensor bulk on the gain performances.

Narrow versus wide gain implants: For similar doping densities and depth of the maximum, narrow gain implants have higher gain than wide ones. Figure 5.4 (*left*) compares the measured gains for two UFSDs, one with shallow-BH and the other with shallow-BL gain implants (Fig. 5.2). The low-diffusion profile generates higher gain than the high-diffusion one, even though, in this specific example, the low-diffusion implant dose is 4% lower. The reason for this difference is that the field increases in the gain implant and reaches its maximum in the gap region. Wide implants have a gap region that is shorter due to the gain implant tails.

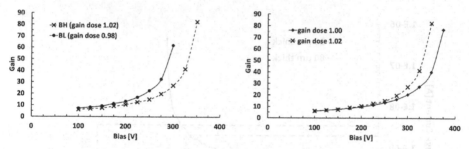

Figure 5.4 *Left*: comparison between gain curves of UFSDs with a shallow-BL and a shallow-BH gain layer. The low-diffusion gain implant is 4% less doped than the high-diffusion implant. *Right*: comparison between gain curves of UFSDs with an equal gain layer design (shallow-BH), but with a 2% difference in dose. The steeper curve belongs to the device with higher dose.

Deep versus shallow gain implants: For equal doping implant, a UFSD with a deep gain implant has a higher gain than a device with a shallow implant. This follows directly from the discussion in Section 2.1.1: equal doping generates an equal field, so the electric field is identical in both designs. Therefore, since the mean drift distance λ is the same in both cases, a wider gain layer has a higher gain since it contains more multiplication steps. A straightforward consequence is that a given gain value is obtained with a lower implant dose in deep gain implants than in shallow ones, this effect is shown on the *left* panel of Fig. 3.12. A less obvious fact is that the position of the gain implant has consequences on the gain implant radiation resistance since acceptor removal has a stronger effect in less doped implants (see Section 6.2). However, this fact does not immediately reflect in a worse temporal resolution: the bias voltage has a more substantial gain recovery capability in UFSDs with a deep gain implant, compensating, at least partially, the faster acceptor removal (see Section 2.10).

High versus low gain implant doping densities: The dose of the gain implant determines the UFSD internal gain: at the same external bias, a higher dose generates a higher gain. The gain curves in Fig. 5.4 (*right*) are of two UFSDs with same gain implant depth and width, but with a 2% different in gain dose. The two gain curves are similar in shape but shifted by about 24 V. This shift, ~ 12 V each % of doping, is a fairly constant number, regardless of the design specifics.

Thin versus thick bulk: The thickness of the active bulk has a direct impact on the gain curve. If the same type of gain layer is present in two UFSDs of different active thicknesses, d_1 and d_2, the gain(V) characteristics will be shifted by about the ratio d_1/d_2. Figure 5.5 shows the $I(V)$ curves of two sensors, one 50 μm and the other 80 μm-thick. Two points on the $I(V)$ curves at the same gain (i.e., two points such that $I_2 = I_1 \cdot 80/50$) are shifted in voltage by about the ratio of thicknesses $440/650 = 0.67 \sim 50/80 = 0.62$. The thickness of the sensor also influences the curvature of the $G(V)$ curve: thick sensors have milder curvature than thinner ones.

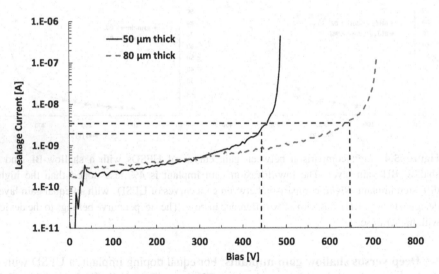

Figure 5.5 Comparison between $I(V)$ curves of two UFSDs with different active thickness (50 μm and 80 μm) and equal gain layer design (deep-BH).

5.1.3 CARBONATED GAIN IMPLANT

The implantation of carbon in the volume of the gain implant aims to improve the radiation resistance of the sensor. As discussed in Section 2.10, the interstitial defects created by irradiation, Si_i, form acceptor-defect complexes deactivating acceptor atoms. This leads to a reduction of the gain implant active density. Since the carbon atoms also form complexes with Si_i, their presence introduces a competitive process that reduces the probability of generating acceptor-defect complexes.

The effects of carbon co-implantation have been studied in deep and shallow gain implants as a function of the carbon dose (UFSD2, UFSD3, and UFSD3.2). In these studies, the carbon dose C_{dose} is quoted relative to the boron density, in unit of the lower carbon dose implemented in the UFSD3 production:

$$C_{\text{dose}} = \frac{(C_{\text{doping}}/B_{\text{doping}})}{(C_{\text{doping}}/B_{\text{doping}})_{UFSD3}}, \qquad (5.5)$$

where B_{doping} and C_{doping} indicate the boron density and the carbon density, respectively.

Table 5.3 reports the carbon doses that have been investigated.

Carbon is activated at a higher temperature than boron, so two activation schemes are possible: (i) implant and activate carbon first and then implant and activate boron (scheme called CHBL), or (ii) implant carbon and boron and then activate them together (called CBL or CBH). In our studies, shallow-carbonated gain implants follow scheme (i) while deep-carbonated gain implants follow either of schemes (ii), as reported in Fig. 5.6.

Table 5.3

Overview of Carbon Doses (Normalized Ratios $C_{\text{doping}}/B_{\text{doping}}$) Which Have Been Investigated in Co-implanted Shallow and Deep Gain Implants

C dose [a.u.]	0.4	0.6	0.8	1	2	3	5	10
Shallow	x			x	x	x	x	x	x
Deep		x		x				

The co-implantation of carbon in the gain implants changes the properties of sensors in several ways. Here, three of the most important are presented.

Leakage current increase: It is known that the presence of carbon increases the detector leakage current. In this specific case, carbon is implanted in a small fraction of the total volume, reducing this effect. The current increase is a function of the carbon dose: Fig. 5.7 shows the leakage currents of UFSDs enriched with carbon doses up to 10 a.u. It has been experimentally proven that this higher leakage current does not degrade the temporal performances of UFSDs. The leakage current rises very rapidly for carbon doses up to about 2, and then it becomes almost constant (the small drop visible on the plot is due to lower gains).

Reduction of active acceptors: Carbon co-implantation reduces the amount of active acceptors. Considering a given gain implant density and activation process, the fraction of activated acceptors decreases as a function of increasing co-implanted carbon dose. This effect, called carbon-boron inactivation, has been studied by comparing V_{GL} as a function of the implanted carbon dose. The *top* side of Fig. 5.8 shows the $C(V)$ curves for 5 different configurations: no carbon, 1, 2, 3, and 5 units of carbon dose. The plot refers to shallow-BL gain implants.

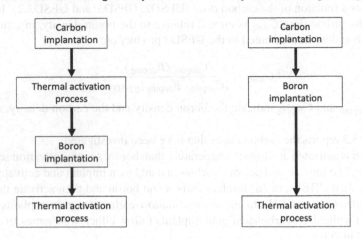

Figure 5.6 Flow charts of the implantation and activation process for shallow- and deep-carbonated gain implants.

The active fraction of the gain implant can be estimated as:

$$F_{GL-Implant} = V_{GL}(carbon)/V_{GL}(no-carbon), \qquad (5.6)$$

where $V_{GL}(carbon)$ and $V_{GL}(no-carbon)$ (see Eq. (5.4)) are the depletion voltages of the gain layers in carbonated and not carbonated UFSDs, respectively. The active fraction of the gain implant decreases as a function of the carbon dose and its decreasing trend depends on the activation schemes used, as shown on the *bottom* side of Fig. 5.8. In CHBL gain implants, $F_{GL-Implant}$ decreases linearly in a carbon dose range 1–3; for carbon doses above 3, the carbon-boron inactivation slows down. For carbon doses below ∼ 0.8, the boron deactivation is almost absent, suggesting a threshold mechanism of the carbon-boron inactivation. From the fit in Fig. 5.8 (*bottom*), the threshold carbon dose is estimated to be 0.76. In CBL(H) gain implants, the carbon-boron inactivation is stronger than in CHBL gain implants and there seems to be no threshold effect. More investigations are needed to further understand this difference.

The carbon-boron inactivation has an effect on the gain of UFSDs: the carbon dose 1 in CHBL-implants causes the deactivation of ∼ 2% of the gain implant, resulting in a gain curve shift of ∼ 25 V towards higher bias. Figure 5.9 shows how the co-implantation of a unit of carbon dose in a CHBL gain implant has the same effect on the gain curve as the decrease of 2% of the initial gain implant doping.

Reduction of the gain implant diffusion: The co-implantation of carbon into the gain implant decreases the lateral diffusion of acceptors during the thermal activation process, especially for boron-L implants. Figure 5.10 compares the profiles of a carbonated (carbon dose 1) and not carbonated shallow-BL gain implant profiles: the carbonated gain implant is ∼ 10% narrower and higher.

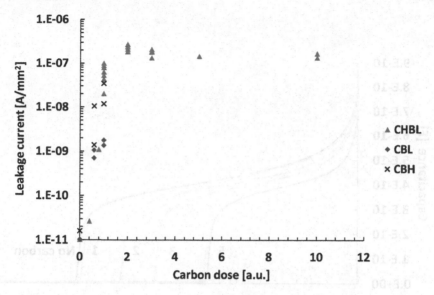

Figure 5.7 Leakage currents for several UFSDs biased at 100 V, as a function of the dose of co-implanted carbon (a.u.).

5.1.4 THE EFFECTS OF TEMPERATURE ON THE GAIN LAYER PERFORMANCES

The temperature at which a UFSD operates has a strong impact on its gain. Combining Eq. (2.2) with Eq. (2.3), the electrons and holes ionization coefficients α can be expressed as:

$$\alpha_{n,p}(\mathscr{E}) = A_{n,p}e^{-\frac{C_{n,p}+D_{n,p}T}{\mathscr{E}}}. \qquad (5.7)$$

The mean drift distance needed to achieve multiplication becomes shorter as the temperature decreases (see Fig. 2.2), increasing the gain for constant bias conditions. Figure 5.11 shows how the gain changes lowering the temperature by 50 K, from 300 K to 250 K, for a 45 μm-thick UFSD with a deep gain implant (*left*) or a 45 μm-thick UFSD with a shallow gain implant (*right*). In order to keep the gain constant, for example gain $G = 20$, as the temperature decreases from 300 K to 250 K, the bias voltage needs to be lowered. Since for a deep implant dG/dV is larger than for a shallow implant, see Section 2.10, the voltage decrease is lower. This effect, compounded with the different thickness, leads to very different bias voltages decrease: $\Delta V_{45\,\mu m,deep} \sim 50$ V instead of $\Delta V_{55\,\mu m,shallow} \sim 90$ V, yielding to $\Delta V_{45\,\mu m,deep} \sim 1$ V/K vs $\Delta V_{55\,\mu m,shallow} \sim 2$ V/K.

Using the data reported in Fig. 5.11, the value of $D_{n,p}$ has been determined to be $D_{n,p} = 990$ V/K. With this value, the WF2 simulations show an excellent agreement with the data.

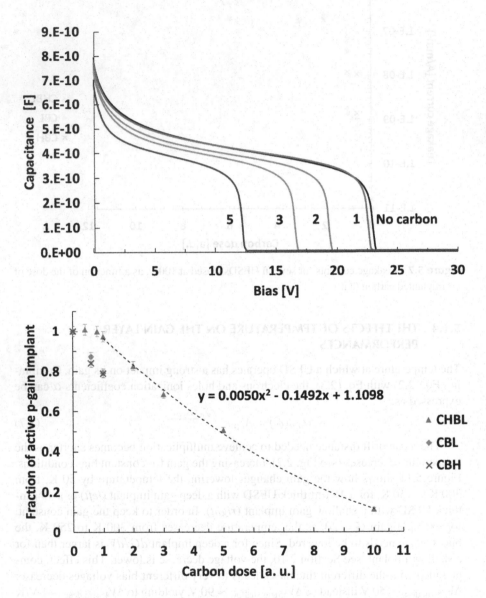

Figure 5.8 *Top*: $C(V)$ curves for 5 different types of gain layer: no carbon co-implantation, 1, 2, 3, and 5 units of carbon dose. *Bottom*: fraction of active gain implant as a function of the carbon dose for shallow-BL (CHBL diffusion process), deep-BL (CBL diffusion process) and deep-BH (CBH diffusion process) gain layers.

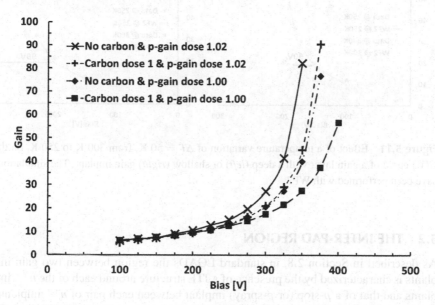

Figure 5.9 Gain as a function of the bias voltage for different combination of carbon and boron doses in the gain implant. The co-implantation of a unit of carbon dose in a CHBL gain implant reduces the gain by the same amount of a decrease of 2% of the initial gain implant doping.

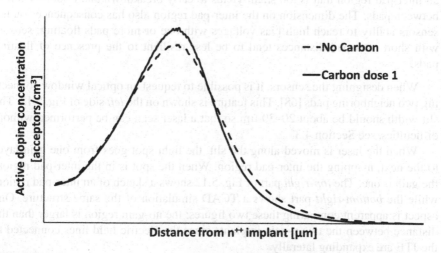

Figure 5.10 Doping profiles of a not carbonated and a carbonated (carbon dose 1) shallow-BL gain implant.

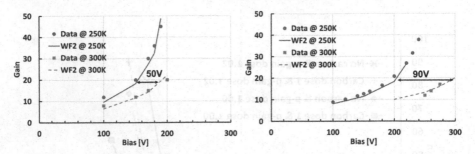

Figure 5.11 Effect of a temperature variation of $\Delta T = 50$ K, from 300 K to 250 K, on the $G(V)$ curve of a gain layer with a deep (*left*) or shallow (*right*) gain implant. The simulations have been performed with WF2.

5.2   THE INTER-PAD REGION

As described in Section 2.8, in standard LGADs the region between two gain implants is characterized by the presence of a JTE structure around each of the n^{++} implants and that of a p-stop (or p-spray) implant between each pair of n^{++} implants. Several designs of the inter-pad region, differing in the width of the JTE and the number, width, and doping of the p-stop implant, have been developed by UFSDs producers. The extent of the inter-pad region is a crucial parameter in the design of multi-pads devices since the arrival time of the particles hitting in this region cannot be determined accurately. On the one hand, the inter-pad region needs to be minimized to increase as much as possible the detector fill factor, but, on the other hand, an inter-pad region that is too small yields to early breakdown and to non-isolation between pads. The dimension on the inter-pad region also has consequences on the sensors ability to reach high bias voltages with one or more pads floating: sensors with short inter-pad distances tend to be less resilient to the presence of floating pads.

When designing the sensors, it is possible to request an optical window connecting two neighboring pads [68]. This feature is shown on the *left* side of Fig. 5.12. The slit width should be about 20–30 μm so that a laser scan can be performed without difficulties, see Section 4.3.

When the laser is moved along the slit, the light spot goes from one gain layer to the next, mapping the inter-pad region. When the spot is in the inter-pad region, the gain is one. The *top-right* part of Fig. 5.12 shows a sketch of an inter-pad region, while the *bottom-right* part shows a TCAD simulation of the same structure. One aspect is apparent comparing these two figures: the no-gain region is larger than the distance between the two p-gain implants since the electric field lines connected to the JTE are expanding laterally.

The result of a typical inter-pad measurement is shown in Fig. 5.13: the signals seen on each of the two adjacent pads are plotted on the y-axis as a function of the

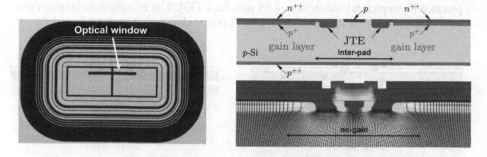

Figure 5.12 *Left*: an example of optical window connecting two neighboring pads. *Right*: a sketch of an inter-pad region (*top*) in an LGAD and its TCAD simulation (*bottom*).

laser position. When the laser is positioned at one side of the optical window, the signal is visible only in one pad. Then, as the laser approaches the inter-pad region, the second pad starts seeing a signal too. When the laser is above the inter-pad region, the signal is small, gain = 1, in both pads. As the laser moves to the second pad, the signal in the first pad disappears. Since the amplitude in one pad is almost zero when the laser is on the other pad, the pads are well isolated. The value of the no-gain distance, 16.7 μm in this particular example, is computed as the 50% distance between the two s-curves, as explained in Section 4.3.

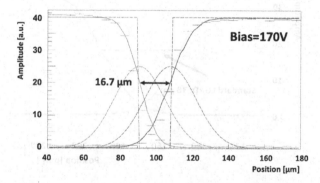

Figure 5.13 Collected charge amplitude (a.u.) read out in two adjacent pads as a function of the laser shot position. The plot shows the measured points and the result of the convolution (solid lines) of a step function (representing the gain implant) and a Gaussian (representing the laser light spot).

5.2.1 TRENCH-ISOLATED LGADS

An alternative technology to isolate neighboring pads uses narrow (about 1 μm) trenches. Trench isolation is extensively used in CMOS imaging sensors [17] and in silicon photomultipliers [23]. When applied to the LGADs design, the trenches, dug with a deep reactive ion etching technique and filled with silicon oxide, replace the JTE and *p*-stop implants. The first Trench-Isolated-LGADs (TI-LGADs) have been produced on epitaxial substrates ~ 55 μm-thick [39]. The comparison between the inter-pad regions of a standard and a TI-LGADs is shown on Fig. 5.14.

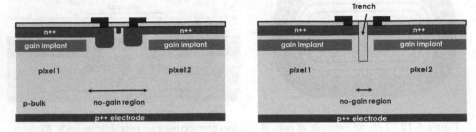

Figure 5.14 Comparison between the standard (*left*) and the trench-isolated (*right*) LGAD designs.

The inter-pad signal profile of a TI-LGAD, shown in Fig. 5.15 [68], proves that this design offers excellent electrical isolation between adjacent pads while significantly reducing the no-gain distance.

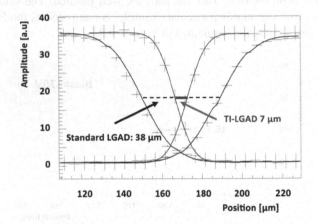

Figure 5.15 Measurement of the inter-pad region of a T2 TI-LGAD and of a standard LGAD [68].

Two different trench designs have been implemented in the first TI-LGADs production. The first one has a trench grid between pads (version T1, Fig. 5.16 *left*), while in the second one, each pad is surrounded by an independent trench ring (version T2, Fig. 5.16 *right*).

The nominal distance between the gain implants in the T1 and T2 designs is ~ 4 μm and ~ 6 μm, respectively. Studies on both designs show excellent $I(V)$ characteristics, high breakdown voltage, and the absence of additional noise contributions. The successful production of TI-LGAD prototypes paves the way to the production of UFSD pixel sensors with small pitch and good fill factor.

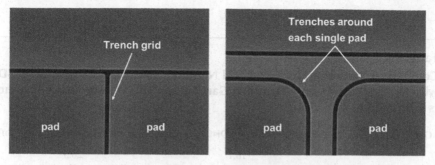

Figure 5.16 SEM imagines of TI-LGAD trenches before the filling process. *Left*: 1-trench grid design (T1). *Right*: trench ring design (T2).

5.2.2 MEASUREMENT OF THE NO-GAIN DISTANCE

The no-gain distances of standard shallow low (high) diffusion, deep low (high) diffusion, and shallow high diffusion TI-LGAD gain layer designs are shown in Table 5.4. The first column reports the inter-pad type, the second column the measured no-gain distance, and the third one the nominal distance between gain layers.

As predicted by simulation, the measured no-gain distance is larger than the nominal one. This difference is about 10–15 µm. Notably, the TI design is able to shrink the no-gain distance to less than 10 µm.

As shown in Fig. 5.12, a part of the no-gain distance is due to the lateral bending of the electric field lines. How much these lines are bent is a strong function of the bias voltage at low bias values, while there is almost no change at high bias values. The dependence of the no-gain distance upon the bias voltage for two different types of inter-pad designs (type with nominal distance B and C) is shown in Fig. 5.17. The bias ranges explored in the plot are centered around the operating voltage of each sensors. Both types show a small decrease of the no-gain distance as the bias voltage rises: the no-gain width varies by about 1 µm for a bias variation of about 35–45 V.

5.2.3 EFFECTS OF THE p-stop IMPLANT DESIGN ON UFSD PERFORMANCES

Small inter-pad designs improve the sensor fill factor; however, they lower its capability to hold high bias voltages. The p-stop structure is floating, so it floats to a potential between that of the n^{++} implant and the bias (normally much closer to n^{++} than to the bias value). There is, therefore, a strong electric field between the p-stop and the n^{++} pad: the shorter this distance, the higher the field. To study this effect, UFSDs with four different inter-pad nominal distances, A = 11 µm, B = 20.5 µm, C = 31 µm, and D = 41 µm, had been manufactured and tested. The $I(V)$ characteristics of these 4 designs are shown in Fig. 5.18. This plot indicates two interesting features: (i) larger inter-pad distances lead to higher breakdown voltages, and (ii) only the design with the widest inter-pad distance (D) has a breakdown due to gain. This can be recognized by the different shape of the $I(V)$ characteristics: A, B, and C have sharp current increases, while D has a smooth exponential behavior.

Table 5.4

Comparison Between Measured and Nominal No-gain Distances for UFSDs with Shallow-L/H, Deep-L/H, and TI Gain Layers. The Measurement Accuracy Is Estimated to be ±2 μm

Gain Layer Type	Measured No-gain Distance[a] [μm]	Nominal No-gain Distance [μm]
Shallow-L	16.7	20.5
	30.4	31
	35	23.5
	38.3	41
	68	49
Shallow-H	31	16
	31	20.5
	32.5	20.5
	62	49
Deep-L	39	23.5
	65	49
Deep-H	42.6	23.5
	71	49
Shallow-H TI-LGAD T1	9	4
Shallow-H TI-LGAD T2	7	6

[a] Measurements performed at the operating voltage of each sensor.

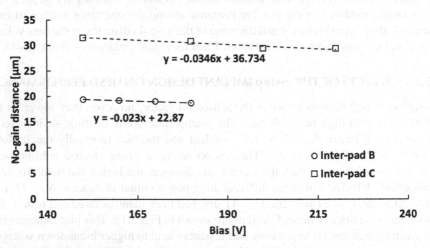

Figure 5.17 No-gain distance as a function of the bias voltage for two different inter-pad designs (layout B and layout C).

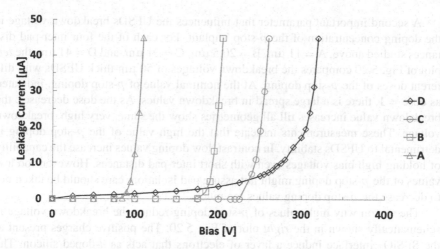

Figure 5.18 $I(V)$ curves of UFSDs with four different inter-pad nominal distances: A = 11 μm, B = 20.5 μm, C = 31 μm, and D = 41 μm.

The origin of the premature breakdown was further investigated using the TCT and the CCD-camera setups. A x-y TCT scan in the region between pads, when biasing the detector close to breakdown, allows detecting the area of high electric fields since the signal in those locations becomes larger.

The CCD-camera images of the same sensor biased at 200 V, 250 V and $\sim$ 260 V, see Fig. 5.19, confirm the TCT measurements. When the bias voltage is set at 260 V, bright spots appear at the corners of the pads. The same studies, performed on the sensor with the largest inter-pad, D = 41 μm, did not reveal any hot spots by either the TCT or the CCD-camera setups. This confirms what was inferred looking at the shape of the $I(V)$ characteristics: the sensor with the largest inter-pad reaches breakdown due to gain and not due to high electric fields on the sensor surface.

Figure 5.19 CCD-camera images of a UFSD sensor with an inter-pad distance of 31 μm at three bias voltages, 200 V, 250 V and 260 V. The hot spots (white spots) appear in the corner of the pads when the bias approaches the breakdown voltage.

A second important parameter that influences the UFSDs breakdown voltage is the doping concentration of the p-stop implant. For each of the four inter-pad distances studied above, A = 11 μm, B = 20.5 μm, C = 31 μm, and D = 41 μm, the *left* plot of Fig. 5.20 compares the breakdown voltages of 50 μm-thick UFSDs with different doses of the p-stop doping. At the nominal value of p-stop doping, indicated as dose = 1, there is a large spread in breakdown values. As the dose decreases, the breakdown value increases till all geometries show the same, very high, breakdown voltage. These measurements indicate that the high value of the p-stop doping is detrimental to UFSDs stability. In contrast, low doping values increase the capability of holding high bias voltages even with short inter-pad distances. However, too low values of the p-stop doping might not assure pad isolation: care should be taken not to decrease the p-stop doping values too much.

The reason why high values of p-stop doping reduce the breakdown voltage is schematically shown in the *right* plot of Fig. 5.20. The positive charges present at the Si-SiO$_2$ interface induce a layer of electrons that acts as n-doped silicon. The value of the p-stop doping determines how abrupt the pn junction is and how high the electric field is.

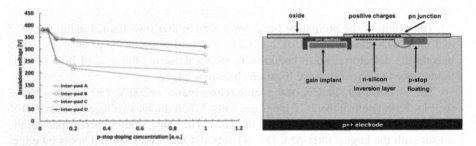

Figure 5.20 *Left*: breakdown voltage as a function of the p-stop doping concentration of four 50 μm-thick UFSD with different inter-pad layouts. *Right*: schematic representation of the inter-pad region where the positive charges forming in the oxide are shown.

The p-stop doping concentration also has an important effect on the breakdown voltage of pixellated UFSDs when one, or more, of its pads is floating (for example, due to a broken wire-bond or a missing bump-bond). In this condition, the floating pad creates a large electric field between itself and the surrounding pads. The consequences on the breakdown voltage are shown in Fig. 5.21 for a UFSD geometry with 2 $\times$ 2 pads. For a high p-stop doping concentration, solid lines, the breakdown voltage decreases as a function of the number of floating pads for every inter-pad distance. The reason, as above, is the high electric field at the inversion layer/p-stop junction. Conversely, for a low p-stop dose, dashed lines, the breakdown voltage is stable even when half of the UFSD surface is not grounded.

Lastly, a too high p-stop doping concentration is also responsible for the appearance of micro discharges [81] that increase the noise and degrade the UFSD performances. An example of micro discharges is shown in Fig. 5.22: the *top*

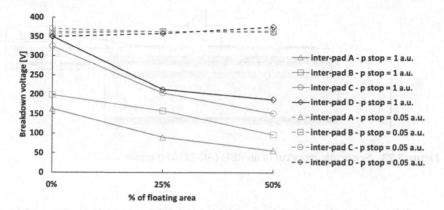

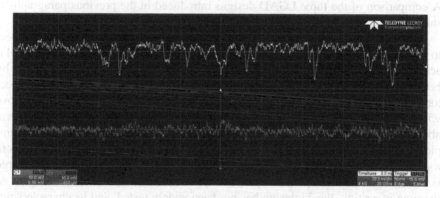

Figure 5.21 Dependence of the breakdown voltage upon the percentage of UFSD pads not connected to ground, for four different inter-pad layouts (A, B, C, and D) and two different p-stop doses (1 and 0.05 a.u.).

(*bottom*) panel shows the oscilloscope trace of a UFSD with (without) micro discharges. This effect appears well below the sensors breakdown voltage and makes impossible the efficient use of the sensor.

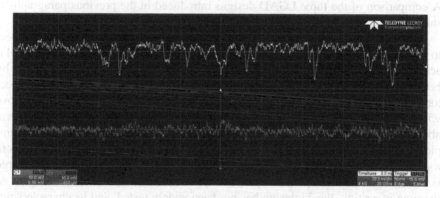

Figure 5.22 Comparison between an oscilloscope trace affected by micro discharges (*top*) and one without (*bottom*).

5.2.4 RESISTIVE AC-COUPLED SILICON DETECTORS (RSD)

Resistive AC-Coupled silicon detectors, also often called AC-LGAD, are an evolution of the LGAD design [61] aimed at eliminating the no-gain area. RSDs, as explained in Section 3.6, are n-in-p sensors, with a continuous gain layer, a resistive n^+ implant, and a thin dielectric layer for AC coupled read-out. An RSD cross cut is shown in Fig. 5.23.

The size of the AC metal pads determines the read-out segmentation. It can be adjusted to any geometry by simply changing two production masks (metal etching and

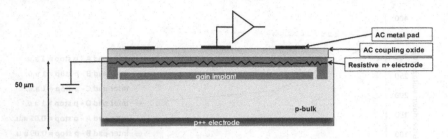

Figure 5.23 Schematic cross cut of an RSD (AC-LGAD) sensor.

overglass), leaving the rest of the sensor identical. The goal of the resistive n^+ layer is to keep the signal localized, to reduce the capacitance seen by the read-out pad, and to induce the AC signal on the metal pad. The role of the n^+ layer is somewhat equivalent to that of the graphite layer in Resistive Plate Chambers (RPC)[1]. Given the continuous gain implant, RSDs offer 100% fill factors. RSDs have been successfully produced [52, 54, 32] and tested [58, 14].

5.2.5 SUMMARY OF INTER-PAD DESIGNS

A comparison of the three LGAD designs introduced in the previous paragraphs is shown in Fig. 5.24.

The JTE/p-stop design, *top* drawing, has a much larger no-gain area with respect to the other two, however, it is the design that so far has been tested the most, and it achieves good performances even after high fluences. This design has been chosen by both the ATLAS and the CMS collaborations for their timing layers. In these detectors, the pad size is relatively large, 1.3×1.3 mm^2, and the fill factor remains above 95%. The TI-LGAD design, *middle* drawing, offers a much smaller no-gain area, about 5 μm, allowing the possibility to design matrices with small pads, for example, 100×100 μm^2, still achieving a fill factor of about 90%. The TI design leads to a sensor with the same working principle of the JTE/p-stop design, the signal charge is collected only in one pad, so these two types of LGAD can be interchanged. At the time of writing, the TI design has not been widely tested, and its characteristics after irradiation have not been studied. The RSD design, *bottom* drawing, differs significantly from the other two types of LGAD. This design leads to a 100% fill factor, and it offers a much simpler and solid layout as the electric field is not interrupted. However, the principle of operation of RSDs is different: the signal is always shared among several pads, and the read-out electronics need to be designed to accommodate this feature. Also, since the signal of every hit is seen on several pads, RSD works best in experiments with low occupancy. The radiation hardness of the RSD design is presently under evaluation.

---

[1]RPCs consist of two electrode plates, both made from a resistive material with metal contacts on the outer part, separated by a thin layer of gas. When a charged particle ionizes the gas molecules, the electrons/ions travel towards the electrodes, and AC-coupled signals are seen on the metal contacts.

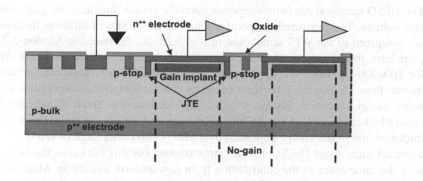

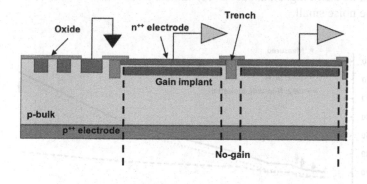

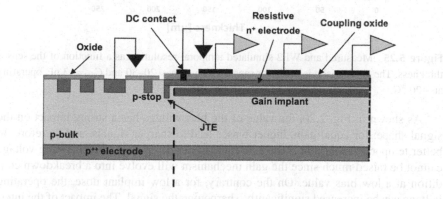

Figure 5.24 Cross sections, not to scale, of a standard LGAD, with JTE/*p*-stop design (*top*), a Trench Isolated LGAD (*middle*) and an RSD (*bottom*). Vertical dashed lines indicate a sharp separation between full-gain and no-gain region [50].

5.3   TEMPORAL RESOLUTION

The UFSD temporal resolution depends upon the sensor thickness, its gain, and the bias voltage. The measured temporal resolutions of UFSDs of different thicknesses are compared to the WF2 simulation in Fig. 5.25. As explained in Section 2.7, the jitter term monotonically decreases as a function of the sensor thickness. While for 100–300 μm-thick UFSD the jitter term dominates the resolution, for sensors thinner than 75 μm, the jitter term becomes comparable to the contribution due to non-uniform ionization. Sensors with an active thickness of about 50 μm have been chosen by both the ATLAS and CMS collaborations for their timing layers. With this thickness, the initial charge deposited by a MIP is still rather large (about 0.5 fC), so a limited gain, about 15–20, provides large signals. For this thickness, the jitter term is of the same order as the contribution from non-uniform ionization. Much thinner sensors, for example 25 μm-thick UFSD, provide very sharp signals; however, the gain should be much higher, about 30–40, and the electronics should be faster while keeping the noise small.

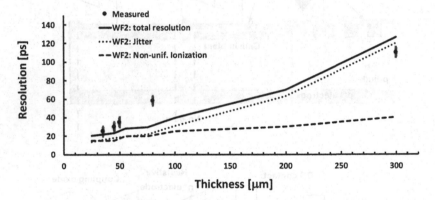

Figure 5.25 Measured and WF2 simulated temporal resolution as a function of the sensor thickness. The comparison is made for sensors with gain $\sim$ 20–30 and $C_{det} = 3$ pF, operating at $-20\,°C$.

As shown in Fig. 2.20, the value of the bias voltage has a strong impact on the signal shape: for equal gain, higher biases lead to sharper signals, and therefore, to better temporal resolution. If the gain implant density is high, the operating voltage cannot be raised much since the gain mechanism will evolve into a breakdown condition at a low bias value. On the contrary, for a low implant dose, the operating voltage can be increased significantly, sharpening the signal. The impact of the interplay between the gain dose and the bias voltage on the temporal resolution is shown in Fig. 5.26 [68]. In the gain-bias plane, the plot reports the temporal resolutions for several $\sim$ 50 μm-thick UFSDs with different gain layer designs. The low-bias high-gain condition leads to a poor temporal resolution, about 45–50 ps, even with a very large gain, gain $\sim$ 30–40. As the bias increases, the best possible temporal resolution, 30–35 ps (solid squares), is achieved first with a large gain, and then even at a gain value of about 15. Overall, the clear trend is that high gain is not enough to assure an excellent temporal resolution: the electric field in the bulk region should be higher than 20–30 kV/cm.

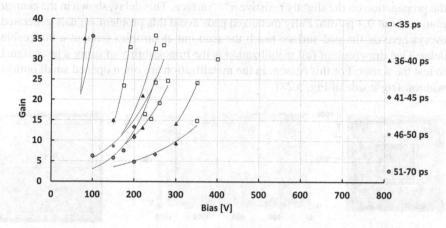

Figure 5.26 Temporal resolutions in the gain-bias plane for several $\sim$ 50 μm-thick UFSDs.

In most cases, the temporal resolution is computed either using the CFD or the ToT techniques (Section 2.6.1). In both cases, the value at which the discriminator threshold is positioned needs to be chosen with care. Figure 5.27 illustrates the effect of the CFD value on the temporal resolution. At low CFD values, the resolution is governed by the jitter term. For this reason, here the gain also has a strong effect. A CFD value around 0.2–0.3 minimizes the resolution, while at high CFD values, the resolution worsens.

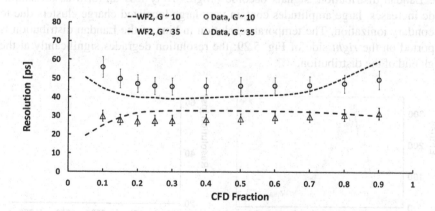

Figure 5.27 Data and WF2 simulation of the temporal resolution of a 50 μm-thick UFSD as a function of the CFD value for two different gains.

Another feature that has a significant impact on the temporal resolution is the absence of full metalization on the pad (see Section 2.5.3). This effect is illustrated in Fig. 5.28: if the pad is not entirely covered with metal, signals suffer a delay due to the propagation on the slightly resistive n^{++} surface. This delay, shown in the *central* plot, is about 0.4 ps/μm. Fully metalized pads avoid this problem: signals generated everywhere on the pad surface reach the read-out electronics without a noticeable delay. One drawback of full metallization is the impossibility of using a laser signal to test the sensor. For this reason, in the metallization are often opened small optical window (*right* side of Fig. 5.28).

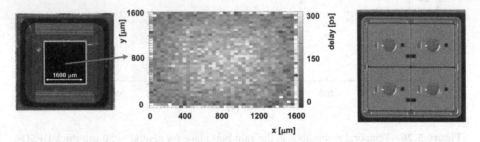

Figure 5.28 *Left*: 2×2 mm^2 50 μm-thick UFSD without metal in the central area. *Center*: the signal time of arrival as a function of position. *Right*: a fully metallized 4-pad 50 μm-thick UFSD (HPK2) with optical windows (seen as dark opening).

The effect of non-uniform ionization determines the lower limit of the UFSDs temporal resolution (see Eq. (2.12)). The effects of non-uniform ionization can be seen studying the temporal resolution, for a given gain, in bins of amplitude, i.e., computing the resolution in slices of the Landau distribution, see Fig. 5.29. Within the Landau distribution, signals become progressively less uniform as the amplitude increases: large amplitudes contain very large localized charge clusters due to secondary ionization. The temporal resolution in bins of the Landau distribution is reported on the *right* side of Fig. 5.29: the resolution degrades significantly at the high end of the distribution.

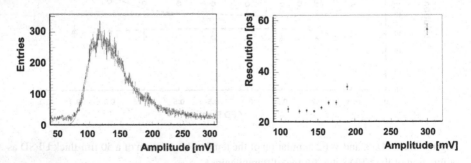

Figure 5.29 Temporal resolution of a 50 μm-thick UFSD with a gain of about 20 (*right*), as a function of the signal amplitude in bins the Landau distribution (*left*).

5.4 YIELD AND UNIFORMITY OF A LARGE UFSD PRODUCTION

The ATLAS and CMS experiments have selected UFSDs for their respective timing detectors [9, 8]. The final design of the UFSD sensor for ATLAS (CMS) will have 450 (512) 1.3×1.3 mm^2 pads, for a total area of about 2×2 cm^2. Each of the timing layers will cover a surface of several squared meters. In order to satisfy the ATLAS and CMS requests in terms of performances and cost, the UFSD production must have high yield and very good uniformity.

5.4.1 YIELD AND LEAKAGE CURRENT UNIFORMITY

In the yield study presented here, the pads were classified into three categories: good, noisy, or bad.

1. Good pad: leakage current within 10 times the mode current;
2. Noisy pad: leakage current above 10 times the mode current;
3. Bad pad: the pad does not reach the required minimum voltage, bringing the sensor in premature breakdown.

The study has been performed on 19 wafers of the UFSD3 production, for a total of 152 UFSDs. Each sensor has 96 1×3 mm^2 pads, arranged in a 4×24 matrix of total area of ~ 3.25 cm^2. The total number of pads in this study is about 15,000. For each sensor, the $I(V)$ characteristic, the breakdown voltage, and the leakage current of each pad have been measured. The measurements set-up to perform this characterization is described in Section 4.1.3.

Overall, $\sim 0.1\%$ pads have been classified as noisy and 0% as bad. Restricting the analysis to those wafers whose gain matched the ATLAS or CMS requests, the noisy pads percentage increases to $\sim 0.2\%$. The complete evaluation of this testing campaign is reported in [57]. While this result shows that the production of UFSDs with many pads is possible, it is worth noticing that a yield of 75% for a sensor with 512 pads requires the good pad probability to be about 0.9995.

5.4.2 GAIN UNIFORMITY

Another critical requirement of a large UFSDs production is the uniformity of gain. This request is especially important for the operation of large sensors, where the bias voltage is common to hundreds of pads. Good gain uniformity is obtained controlling very well the uniformity of the gain implant: the dependence of the collected charge on the gain implant doping is shown in Fig. 5.30. These plots report the relative collected charge as a function of the relative gain implant density. On the *left* side, two 50 μm-thick UFSDs, with a difference of 2% in the gain implant densities, biased at 300 V, differ in the collected charge by about 30%. On the *right* side, two 45 μm-thick UFSDs, with a difference of $\sim12\%$ in the gain implant densities, biased at 250 V, differ in the collected charge by about 75%. On both plots, the results are compared with the WF2 simulations. As the plots indicate, a fairly small change in the doping concentration changes the gain by a large amount. For this reason, good gain uniformity within a sensor, about 10%, requires a gain implant non-uniformity of less than 1%.

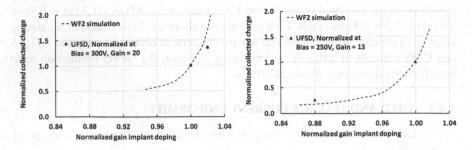

Figure 5.30 Normalized collected charge as a function of the gain implant doping (a.u.). *Left*: two 50 μm-thick UFSDs, with a difference of 2% in the gain implant densities, biased at 300 V. *Right*: two 45 μm-thick UFSDs, with a difference of ~12% in the gain implant densities, biased at 250 V.

A measurement of the gain implant uniformity can be obtained from the $C(V)$ characteristics: the spread of the V_{GL} gives a direct indication of the variability of the gain implant dose, Eq. (5.4). An example of this type of study [68] is shown in Fig. 5.31. In this production, a gain implant non-uniformity of about 2% was measured over the 6-inch wafer surface, which yields to a non-uniformity per sensor of less than 1% (a sensor is much smaller than the wafer).

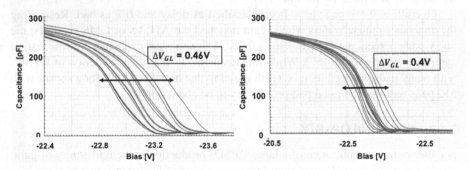

Figure 5.31 Spread of the $C(V)$ measurements on two wafers of the UFSD3 production. The sensors selected for this study were positioned uniformly on the wafer surface.

The depletion voltage of sensors distributed uniformly on four 6-inch wafers (two wafers with a shallow and two with a deep gain implant) was measured, and the results are reported in Table 5.5. The two wafers with a shallow implant have a non-uniformity of ~ 1.9% and ~ 2.2%, while the two wafers with deep implant have a non-uniformity of ~ 2.4% and ~ 3.2%. Combining the wafers with identical implant, a total non-uniformity of ~ 2.5% and ~ 3.7% has been measured for shallow and deep gain implants, respectively.

Table 5.5

Gain Implant Non-uniformity (in %) for Wafers with Gain Implant With Different Depth (HPK1 Production)

Gain Implant Depth	Wafer	$(V_{GL\text{-}max} - V_{GL\text{-}min})/V_{GL\text{-}mean}$ [%]
Shallow	A	2.2
	B	1.9
Deep	A	2.4
	B	3.2

This study also showed a correlation between the pad position on the wafer and the value of V_{GL}: the value of V_{GL} in shallow gain implants slightly increases moving from the centre to the periphery of the wafers. In contrast, in deep gain implants the value of V_{GL} increases from top to bottom (bottom is defined where the wafer circumference has a straight section). The numbers presented in this analysis suggest that shallow implants have a higher uniformity; however, more studies are needed to confirm this effect. Position-dependent non-uniformity is actually a well known effect in the production of silicon devices, it is mostly linked to arise implantation and/or annealing variations over the wafer surface.

Table 5.5

Gain Implant Non-uniformity (in %) for Wafers with Gain Implant With Different Depth (HBK Production).

Gain Implant Depth	Wafer	$V_{center} - V_{edge}/V_{center}$ (%)
Shallow	A	27
	B	10
Deep	A	25
	B	12

This study also showed a correlation between the pad position on the wafer and the value of V_{fd}. The value of V_{fd} in shallow gain implants slightly increases moving from the centre to the periphery of the wafer. In contrast, in deep gain implant the value of V_{fd} increases from top to bottom. There is a trend where the wafer circumference has a straight direction. The numbers presented in this analysis suggest that shallow implants have a higher uniformity; however, more studies are needed to confirm this effect. Position-dependent non-uniformity is actually a well known effect in the production of silicon devices; it is mostly linked to arise implantation and/or annealing variations over the wafer surface.

6 Characterization of Irradiated UFSDs

This chapter focuses on how large fluences of neutrons and protons change the properties of UFSDs. The sensors tested in these studies have an active thickness of about 50 μm, and differ in the gain layer designs and wafer type. They are from the productions HPK1, HPK2, UFSD2, UFSD3, UFSD3.2, CNM1 (see Appendix A).

6.1 IRRADIATION CAMPAIGNS AND HANDLING OF IRRADIATED SENSORS

The results presented in the following paragraphs have been obtained over several years, exposing PIN diodes and UFSDs, without bias, to a broad range of neutrons and protons fluences. The purpose of the study of irradiated devices is to understand how to design UFSDs that are able to function correctly in environments with high radiation levels. Key goals are the development of a parametrization for the acceptor removal mechanism, the investigation of the best designs to maintain the noise level low, and how to extend as much as possible the range of bias voltage used to compensate the effect of acceptor removal.

The irradiations with neutrons have been performed at the TRIGA research reactor at the Jozef Stefan Institute (JSI) in Ljubljana. This facility is used by many groups since its neutron spectrum and flux are very well known [126]. The neutrons energy is about 1 MeV and the irradiation fluence is expressed in 1 MeV neutrons equivalent per cm^2 (n_{eq}/cm^2). For these irradiation campaigns, fluences in the range 0^{14}–$5 \cdot 10^{15}$ n_{eq}/cm^2 have been chosen.

The protons irradiations have been performed at four momenta:

1. 23 MeV/c protons at the KIT irradiation facility in Karlsruhe [91].
2. 70 MeV/c protons at the Cyclotron and Radioisotope Center (CYRIC) at Tohoku University [90].
3. 800 MeV/c protons at the Los Alamos Neutron Science Center (LANSCE) proton accelerator [111].
4. 24 GeV/c proton at the CERN IRRAD beamline [21].

The momenta and the NIEL factors of the four proton irradiation facilities are reported in Table 6.1.

In a real experiment, the irradiation process happens on a timescale of years, while it happens on a timescale of days (or even hours) in a typical irradiation campaign. For this reason, in real experiments part of the damage has time to anneal while this does not happen in devices undergoing the irradiation campaigns. This difference is compensated by annealing the sensors in a thermal chamber before

Table 6.1

Proton Momenta and NIEL Factors for the Irradiation Facilities KIT, CYRIC, LANSCE, and IRRAD

Irradiation facility	Protons momentum [MeV/c]	NIEL [1 MeVn$_{eq}$]
KIT	23	~ 2.5
CYRIC	70	~ 1.5
LANSCE	800	~ 0.71
IRRAD	$24 \cdot 10^3$	~ 0.67

measuring their performances. A typical annealing cycle lasts 80 minutes at 60 °C, which is roughly the time required to anneal short-term radiation-induced defects and corresponds to the time the detectors are kept at room temperature during the yearly maintenance period [37]. Subsequently, the devices are always kept in a cold box at -20 °C. All irradiated UFSDs have undergone the above described annealing cycle.

6.2   STUDY OF THE ACCEPTOR REMOVAL MECHANISM

The study of the acceptor removal mechanism (Section 1.3) on different types of the gain layer is crucial to identify the most radiation-resistant design. Experimentally, it is possible to extract the depletion voltage of the gain layer V_{GL} from the $C(V)$ measurements. Since V_{GL} is proportional to the amount of active doping in the gain implant, Eq. (4.3), the evolution of V_{GL} with fluence measures the deactivation of the gain implant. Figure 6.1 shows a typical evolution of the $C(V)$ curve in irradiated UFSDs, where the decrease of V_{GL} is evident as the fluence increases. The ratio $V_{GL}(\Phi)/V_{GL}(0)$ depends upon the value of the acceptor removal coefficient $c(N_A(0))$:

$$\frac{V_{GL}(\Phi)}{V_{GL}(0)} = \frac{N_A(\Phi)}{N_A(0)} = e^{-c(N_A(0))\Phi}, \qquad (6.1)$$

where $V_{GL}(0)$ and $V_{GL}(\Phi)$ are the gain layer depletion voltages when new or after a fluence Φ, and $N_A(0)$ and $N_A(\Phi)$ the acceptor densities when new or after a fluence Φ. Therefore, the evolution of $V_{GL}(\Phi)/V_{GL}(0)$ with fluence allows extracting the value of $c(N_A(0))$, as shown in Fig. 6.2. The smaller the value of c, the higher the gain implant radiation hardness.

6.2.1   DETERMINATION OF V_{GL} IN IRRADIATED UFSDS

As discussed in Section 4.1.2, a silicon sensor can be modelled either as a C_p-R_p or a C_s-R_s circuit. Since the C_s-R_s circuit does not allow to have a DC current, after irradiation the C_p-R_p is preferred. With this model, V_{GL} can be extracted by either the $1/C^2(V)$ or the $R_p(V)$ curve.

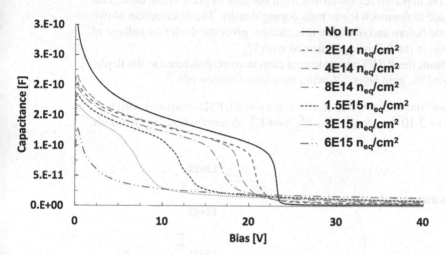

Figure 6.1 $C(V)$ curves of identical UFSDs (shallow-BH gain layer), unirradiated and irradiated with neutron fluences in the range $2 \cdot 10^{14}$–$6 \cdot 10^{15}$ n_{eq}/cm^2.

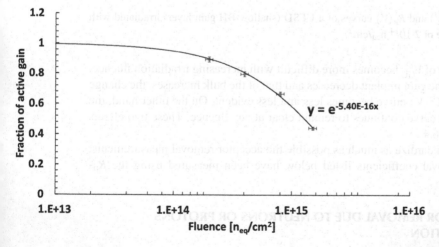

Figure 6.2 Fraction of active gain implant (shallow-BH) as a function of the irradiation fluence (neutron irradiation).

1. **$1/C^2(V)$-method:** the slope of the $1/C^2(V)$ curve has a sharp increase when
   the limit of the depleted region moves from the gain implant to the bulk. This
   increase is due to the much lower bulk doping density. The intersection of two
   linear fits, one before and one after this change, gives the depletion voltage of
   the gain layer, in the following indicated with V_{GL}^C.
2. **$R_p(V)$-method:** the $R_p(V)$ curves have a cusp in correspondence to the depletion voltage of the gain layer, allowing easy identification of V_{GL}^R.

Figure 6.3 shows the $1/C^2(V)$ and $R_p(V)$ curves of a UFSD irradiated with neu-
trons to a fluence of $2 \cdot 10^{14}$ n_{eq}/cm^2. The V_{GL}^C and V_{GL}^R determinations agree within
less than 1 V.

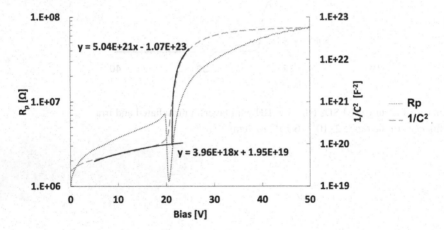

Figure 6.3 $1/C^2(V)$ and $R_p(V)$ curves of a UFSD (shallow-BH gain layer) irradiated with
neutrons to a fluence of $2 \cdot 10^{14}$ n_{eq}/cm^2.

The extraction of V_{GL}^C becomes more difficult with increasing irradiation fluence.
As the doping of the gain implant decreases and that of the bulk increases, the change
of slope in the $1/C^2(V)$ curve becomes less and less evident. On the other hand, the
cusp in the $R_p(V)$ curve continues to remain clear at any fluence. These two effects
are shown in Fig. 6.4.

In order to standardize as much as possible the acceptor removal measurements,
all acceptor removal coefficients listed below have been measured using the R_p-
method.

6.2.2   ACCEPTOR REMOVAL DUE TO NEUTRONS OR PROTONS
IRRADIATION

Nineteen different gain implant types, irradiated with neutrons or protons up to
$1.5 \cdot 10^{15}$ n_{eq}/cm^2, have been characterized to study the acceptor removal mechanism. The gain implants differ by: the implant depth, broad, shallow or deep; the heat
load used in the activation of the dopant, leading to either low diffusion (L) or high

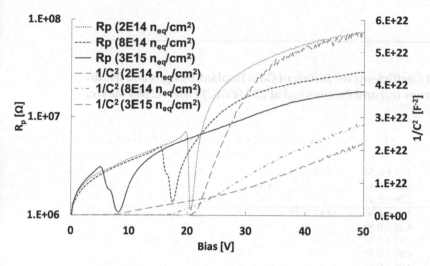

Figure 6.4 $R_p(V)$ and $1/C^2(V)$ curves of UFSDs (shallow-BH gain layer) irradiated with neutrons at fluences 2, 8, and $30 \cdot 10^{14}$ n_{eq}/cm^2.

diffusion (H) designs; the acceptor type, boron or gallium; the amount of co-implanted carbon.

These factors contribute to either increase or decrease the radiation hardness of the gain implant. The corresponding values of the acceptor removal coefficients ($c_{n,p}$) are reported in Table 6.2. By comparing the values of the c_n coefficients, several conclusions can be drawn:

1. The use of gallium instead of boron does not improve the radiation resistance.
2. Carbon co-implantation increases the radiation resistance.
3. There is an optimum range of carbon dose. In arbitrary units, this range is 0.6–1: lower or higher carbon doses yield to lower radiation resistance.
4. A lower heat load, yielding to a less diffused gain implant, improves radiation hardness.
5. Carbonated gain implants activated using CBL or CBH schemes (see Section 5.1.3) are intrinsically more radiation resistance than those activated using CHBL scheme, Fig. 6.5.

Figure 6.6 shows the beneficial effect of carbon co-implantation (carbon dose) for a shallow-BL gain implant: without carbon, 80% of the gain implant is still ctive after a fluence of $\sim 6 \cdot 10^{14}$ n_{eq}/cm^2. In carbonated gain implants, a fluence of $\sim 1.5 \cdot 10^{15}$ n_{eq}/cm^2 is needed to deactivate the same amount of gain implant. In this pecific example, the carbon co-implantation roughly halves the acceptor removal ate.

The value of c_n is plotted as a function of the carbon dose in Fig. 6.7. For clarity, ne value of c_n has been normalized to its value at carbon dose = 1. By increasing ne carbon dose from 0 to about 0.5, the value of c_n decreases rapidly, improving

Table 6.2

Acceptor Removal Coefficients for Different Gain Implant Designs and Irradiation Types (Neutrons (c_n) and Protons (c_p) at 23 MeV/c, 70 MeV/c, 800 MeV/c, and 24 GeV/c)

c [10^{-16} cm²] Gain layer	c_n	$c_{p-23\ \text{MeV/c}}$	$c_{p-70\ \text{MeV/c}}$	$c_{p-800\ \text{MeV/c}}$	$c_{p-24\ \text{GeV/c}}$
Shallow-BL	4.66 ± 0.54	12.79	-	-	-
	3.85 ± 0.47	-	6.91	-	-
	3.79 ± 0.47	-	-	-	-
Shallow-BH	5.40 ± 0.60	14.87	-	-	-
	4.26 ± 0.50	-	-	-	-
	4.47 ± 0.52	-	-	-	-
	4.97 ± 0.55	-	-	-	-
	4.95 ± 0.55	-	-	-	-
Shallow-Ga	7.10 ± 0.72	-	-	-	-
	7.19 ± 0.73	-	-	-	8.35
Deep-BH	5.74 ± 0.61	15.12	-	-	-
Deep-BL	4.18 ± 0.49	-	-	-	-
	5.51 ± 0.63	-	-	-	-
	5.29 ± 0.61	-	-	-	-
	5.21 ± 0.60	-	-	-	-
	5.63 ± 0.63	-	-	-	-
Broad-BH	8.89 ± 0.81	-	-	-	-
Shallow-BL + 0.4C	2.43 ± 0.31	-	-	-	-
Shallow-BL + 0.8C	1.48 ± 0.24	-	-	-	-
Shallow-BL + 1C	1.45 ± 0.28	6.15	3.14	-	-
	1.57 ± 0.29	6.24	-	-	-
	1.91 ± 0.32	-	-	-	-
Shallow-BL + 2C	2.48 ± 0.36	-	-	3.50	-
Shallow-BL + 3C	2.76 ± 0.39	-	-	4.20	-
Shallow-BL + 5C	3.51 ± 0.56	-	-	-	-
Shallow-BH + 1C	2.63 ± 0.38	6.93	-	-	2.25
	2.46 ± 0.36	-	-	-	-
	2.11 ± 0.34	-	-	-	-
Shallow-BH + 2C	3.21 ± 0.43	-	-	-	-
Shallow-BH + 3C	3.53 ± 0.46	-	-	-	-
Deep-BL + 0.6C[a]	2.16 ± 0.28	-	-	-	-
	1.63 ± 0.24	-	-	-	-
Deep-BL + 1C[a]	2.38 ± 0.30	-	-	-	-
	2.06 ± 0.27	-	-	-	-
Deep-BH + 0.6C[a]	1.90 ± 0.26	-	-	-	-
Deep-BH + 1C[a]	2.45 ± 0.30	-	-	-	-
	2.05 ± 0.27	-	-	-	-

[a] Carbon and boron activation in a single process step.

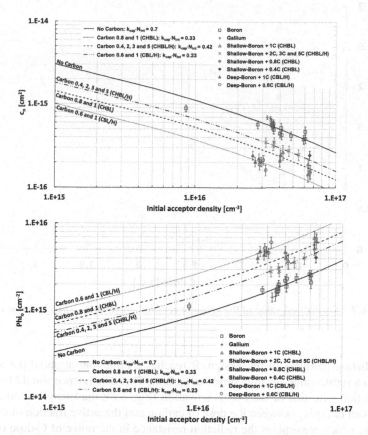

Figure 6.5 Parametrization of the acceptor removal coefficients c_n (*top*) and $\Phi_o = 1/c_n$ (*bottom*) as a function of the initial acceptor density, for carbonated and not-carbonated gain implants.

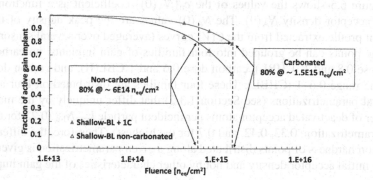

Figure 6.6 Fraction of active gain implant (shallow-BL) as a function of neutron fluence for a carbonated and not-carbonated gain implant (UFSD2).

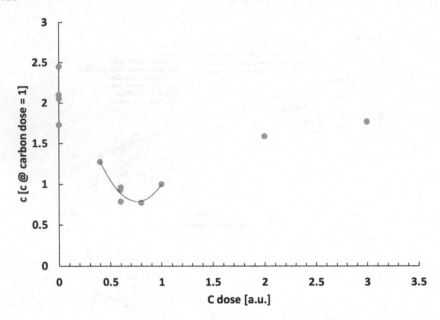

Figure 6.7 Normalized value of the acceptor removal coefficient c_n as a function of carbon dose.

the radiation hardness of the design. It reaches a minimum at about 0.8 and then rises to a stable value for high doses of carbon. This dependence should be seen in conjunction with that of the boron activation (Fig. 5.8): the two plots indicate that there is an interplay between the dose of carbon and the active fraction of the boron implant, which maximizes the radiation resistance in the range of C-dose 0.6–1 (in arbitrary units). Below this range, the radiation resistance decreases rapidly as if there were not enough carbon; above it, a large fraction of boron is not fully activated, and the radiation resistance reaches a plateau.

Figure 6.5 shows the values of the $c_n(N_A(0))$ coefficient as a function of the initial acceptor density $N_A(0)$. The $N_A(0)$ values are the peak density of the gain implant profile extracted from the $C(V)$ curves (averaged over several sensors).

The points can be grouped into four families of gain implants: no carbon, carbon dose 0.8 and 1 (CHBL), carbon dose 0.6 and 1 CBL(H), and carbon dose outside the range 0.6–1 (CHBL). These four families are distributed on four acceptor removal parametrizations (see Section 1.3.2) that differ uniquely by the maximum number of deactivated acceptor atoms per incident particle $k_{cap}N_{Int}$ (0.23 for the lowest parametrization, 0.33, 0.42, and 0.7 for the highest). Therefore, the difference in radiation hardness of points distributed along a given parametrization is given solely by the initial acceptor density and not by other characteristics of the gain implant.

The last four columns of Table 6.2 report the values of the acceptor removal coefficient c_p in proton irradiations. As expected, protons with momentum below

1 GeV/c produce more damage than 1 MeV neutrons, while at high momentum values $c_p \sim c_n$. Figure 6.8 shows c_p as a function of the proton momentum, for two different gain implant designs. The two dashed lines indicate the c_n values for the same gain implant.

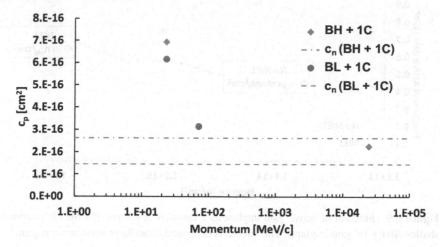

Figure 6.8 Evolution of the proton acceptor removal coefficient c_p as a function of the proton momentum.

The application of the relative NIEL factor to the proton fluence, $\Phi_{n_{eq}} =$ NIEL $\cdot \Phi_p$, has the effect of scaling the x-axis, changing the c_p coefficient as $c_p^{NIEL} = c_p/$NIEL. This effect is shown in Fig. 6.9. In a similar way, the relative NIEL factor measured from the acceptor removal mechanism, $NIEL_{AR}$, can be obtained as the ratio between the proton and the neutrons acceptor removal coefficients: $NIEL_{AR} = c_p/c_n$. This evaluation of the NIEL parameter does not need to be identical to that shown in Fig. 1.6 since the acceptor removal mechanism does not neccessarily scale with the non-ionizing energy loss.

Table 6.3 reports the relative $NIEL_{AR}$ values calculated from the ratio between the protons and neutron acceptor removal coefficients for proton momenta of 23 MeV/c, 70 MeV/c, 800 MeV/c, and 24 GeV/c.

How these values compare with those tabulated in literature [87] is reported in Fig. 6.10. The values of $NIEL_{AR}$ show the same trend with proton momenta of the tabulated NIEL, albeit their values are always higher.

6.3   GAIN IN THE SENSOR BULK

As presented in Section 2.9, four main effects happening in the silicon bulk are degrading the properties of UFSDs after irradiation: (i) increased leakage current, (ii) changed doping profile, (iii) decreased charge collection efficiency, and (iv) reduced mobility. These effects increase linearly with fluence up to about $5 \cdot 10^{15}$ n_{eq}/cm^2, while at higher fluences the dependence upon the fluence becomes logarithmic. The

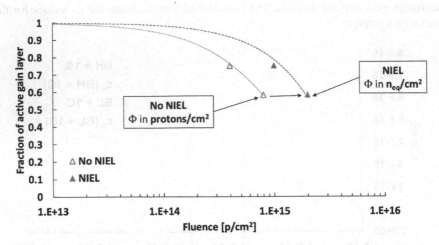

Figure 6.9 Fraction of active gain implant as a function of fluence (23 MeV/c protons on shallow-BH + 1C gain implant). The fluence is expressed either in proton/cm^2 or n$_{eq}$/cm^2.

Table 6.3
Relative NIEL Factors Measured as the Ratio c_p/c_n at Four Different Proton Momenta

Gain layer	$c_{P23\ MeV/c}/c_n$	$c_{P70\ MeV/c}/c_n$	$c_{P800\ MeV/c}/c_n$	$c_{P24\ GeV/c}/c_n$
Shallow-BL	2.75	1.80	-	-
Shallow-BH	2.75	-	-	-
Shallow-Ga	-	-	-	1.16
Deep-BH	2.63	-	-	-
Shallow-BL + 1C	4.24	2.17	-	-
	3.98	-	-	-
Shallow-BL + 2C	-	-	1.41	-
Shallow-BL + 3C	-	-	1.52	-
Shallow-BH + 1C	2.64	-	-	0.86

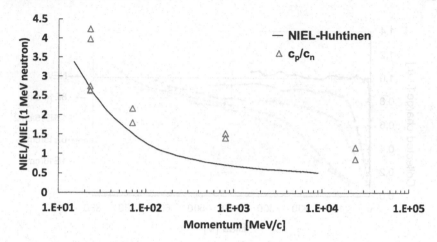

Figure 6.10 Tabulated (solid line) and measured (c_p/c_n) relative NIEL values as a function of the proton momenta.

saturation of radiation damage might extend the range of operation of UFSDs to fluences higher than what initially foreseen (see Chapter 7).

One important effect that might contribute to maintaining a high value of gain in UFSDs is bulk multiplication. The bias voltage can be raised substantially at high enough fluences, generating an electric field in the bulk of about $\mathscr{E} \sim 150$–200 kV/cm. At these values of the electric fields, the mean drift distance is rather long, about 5–10 μm; however, the bulk is several times longer than that, allowing for multiplication to happen. Since the bulk gain is compounded with the gain from the gain layer, even a bulk gain of about two can substantially affect the total gain.

Figure 6.11 shows the collected charge in 55 μm-thick PIN diodes as a function of bias voltage for fluences up to $1 \cdot 10^{16}$ n_{eq}/cm^2. These measurements have been performed at $T = -20\,°C$ to keep as low as possible the bulk leakage current and at two different laser intensities to exclude any dependence on the signal charge density.

The plot shows that for bias voltages in the range 700–800 V, the collected charge increases, indicating bulk multiplication. An interesting feature of this plot is that the gain seems to have the same value for all fluences up to $1 \cdot 10^{16}$ n_{eq}/cm^2.

This effect is further investigated in Fig. 6.12 [66]: the plot at the *top* shows the collected charge in a 45 μm-thick PIN diodes as a function of the bias voltage for three different fluences. In the same plot, the three lines represent the WF2 simulations using the Massey model for avalanche generation.

Since the electric field in the bulk is actually increasing as a function of irradiation (due to the acceptor creation in the bulk), the simulation predicts an increasing gain as a function of fluence. This prediction is not confirmed by the experimental measurements, suggesting that there must be a gain quenching mechanism that reduces the overall gain. An obvious hypothesis is that in an irradiated bulk, the number of scattering centres increases, and the charge carriers do not acquire enough energy to start the impact ionization mechanism. Formally, this mechanism is similar to the

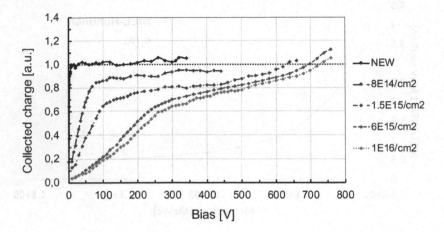

Figure 6.11 Collected charge in a 55 μm-thick PIN diode as a function of the bias voltage for fluences up to $1 \cdot 10^{16}$ n_{eq}/cm^2. The sensor operating temperature is $-20\ °C$.

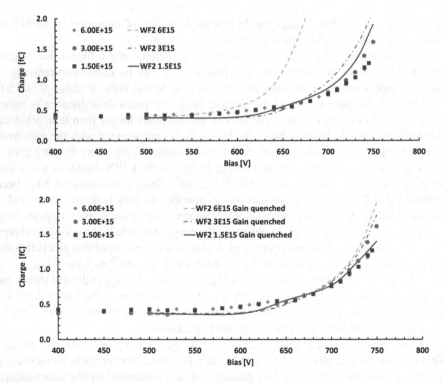

Figure 6.12 *Top*: charge released by a MIP signal in a 45 μm-thick PIN diodes as a function of bias voltage for fluences 1.5, 3, and $6 \cdot 10^{15}$ n_{eq}/cm^2. Overlapped, the predictions from the WF2 simulator. *Bottom*: the same plot with the introduction of a gain quenching term in the simulation [66].

effect of temperature on the gain, see Eq. (5.7), so it can be parametrized adding a fluence-dependent term to the impact ionization coefficients:

$$\alpha_{n,p}(\mathscr{E}) = A_{n,p}e^{-\frac{C_{n,p}+D_{n,p}T+F_{n,p}\Phi}{\mathscr{E}}}. \tag{6.2}$$

Fitting the experimental data, a value of $F_{n,p} = 2 \cdot 10^{-11}$ V·cm is determined. With this extra term, the simulation agrees better with the data, as shown in Fig. 6.12 *bottom*.

6.4 GAIN, NOISE, AND TEMPORAL RESOLUTION OF IRRADIATED UFSD SENSORS

The temporal resolution of irradiated sensors depends on the values of bias, gain, and noise present at a given fluence. In well-designed sensors, even for fluences above 1–$2 \cdot 10^{15}$ n_{eq}/cm^2, the contribution of the sensor noise (shot noise combined with the other noise sources) to the total noise is rather small if the temperature is kept low, $T < -20$ °C, and the pixel size is below 1–2 mm^2. The results shown in the following are obtained using β telescopes, as described in Section 4.4.

6.4.1 WF2 SIMULATION OF THE GAIN OF IRRADIATED UFSDS

The sensor gain depends on the electric field in the gain layer, which is the sum of two components: (i) the field due to the gain implant and (ii) the field due to the external bias, see Section 2.10. The interplay between these two components determines if it is better to have a shallow gain implant - more doped, with smaller acceptor removal and less bias recovery capability - or a deeper gain implant - less doped, with higher acceptor removal and a stronger bias recovery capability. For this reason, the simple comparison of the acceptor removal coefficient values listed in Table 6.2 is not sufficient to select the more radiation-resistant design, and a simulation that considers both aspects is necessary.

Figure 6.13 shows the evolution of the gain $= 20$ bias point with fluence for a broad, a shallow, and a deep gain implant (these last two designs with and without the addition of carbon). The acceptor removal coefficients used in the simulation, chosen from Table 6.2 as indicative of each of the 5 situations, are $c_n^{Broad} = 8.5 \cdot 10^{-16}$ cm^2, $c_n^{Shallow+C} = 1.7 \cdot 10^{-16}$ cm^2, $c_n^{Deep+C} = 2.6 \cdot 10^{-16}$ cm^2, $c_n^{Shallow} = 3.7 \cdot 10^{-16}$ cm^2, and $c_n^{Deep} = 5.5 \cdot 10^{-16}$ cm^2.

The five designs are compared by fixing a common starting condition, gain $= 20$ at bias $= 150$ V, and simulating the necessary ΔV_{bias} to keep the gain $= 20$. The gain layer with the broad design is clearly less radiation resistant, with an increase of about $\Delta V_{bias} = 700$ V at $8 \cdot 10^{14}$ n_{eq}/cm^2. The shallow and deep designs without carbon need a $\Delta V_{bias} = 600$ V after a fluence of $1.5 \cdot 10^{15}$ n_{eq}/cm^2, while carbonated designs require the same bias increase after a fluence 2.5 times higher, at $4 \cdot 10^{15}$ n_{eq}/cm^2. Remarkably, the simulation indicates that for the present values of the c_n coefficients and implantation depths, the interplay between acceptor removal and bias recovery is such that the shallow and deep design perform similarly.

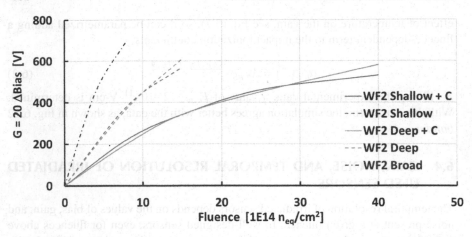

Figure 6.13 WF2 simulation of the evolution of the $G = 20$ bias voltage as a function of fluence, for a broad, a shallow (with and without C), and a deep gain implant (with and without C).

The experimental results on the evolution of the gain $= 20$ bias point with fluence, from pre-irradiation to $\Phi = 1.5 \cdot 10^{15}$ n_{eq}/cm^2, for 4 types of gain layer (deep-BL, shallow-BL + C, deep-BL + C, and deep-BH + C) are shown in Figure 6.14 [122]. The y-axis reports the gain = 20 bias point increase while the x-axis the pre-irradiation gain = 20 bias values. The effect of carbon infusion is very evident: the bias points move by about 300 V for carbonated gain implants while by 550 V for those without carbon. Shallow-BL + C and deep-BH + C behave similarly, $\Delta V \sim 350$ V: shallow-BL + C has a smaller acceptor removal coefficient while deep-BH + C has a stronger bias recovery capability. For this specific production, the two designs yield the same result. Deep-BL + C has a shift that is roughly 70 V smaller, demonstrating that this is presently the most radiation hard design.

6.4.2 TEMPORAL RESOLUTION

There are many publications reporting studies on the temporal resolutions of UFSDs [33, 43, 82, 72], and the results are continuously evolving. The latest results can be reviewed in the presentations available on the web sites of specialized workshops such as RD50 [113], the Tredi Workshop on Advanced Silicon Radiation Detectors, and the Hiroshima Symposium (HSTD). On the more general ground, the present limit for unchanged performances, i.e., the maximum fluence at which is possible to maintain the pre-irradiation temporal resolution, is about $1-2 \cdot 10^{15}$ n_{eq}/cm^2, while, as shown in Fig. 6.13, the most advanced designs have the possibility of reaching $3-4 \cdot 10^{15}$ n_{eq}/cm^2.

In Fig. 6.14 the temporal resolution of several UFSDs, new and irradiated, is displayed in the gain-bias plane, using different symbols to indicate the value of the temporal resolution (separated in five ranges). This plot illustrates well the interplay

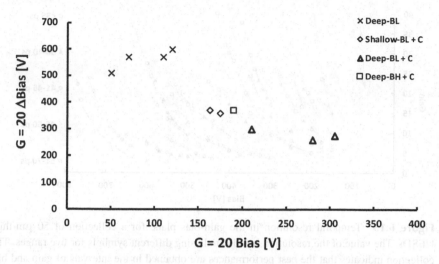

Figure 6.14 Increase of the G = 20 bias point after a fluence of $\Phi = 1.5 \cdot 10^{15}$ n_{eq}/cm^2 for 4 different gain layer design: deep-BL, shallow-BL + C, deep-BL + C, and deep-BH + C (HPK2, UFSD3.2).

of two important contributions to the temporal resolution: the $\mathscr{E}$ in the bulk to saturate the velocity of the charge carriers, and the total gain, to minimize the jitter. The plot presents a collection of results for sensors from various productions (UFSD2, UFSD3, UFSD3.2, HPK1, HPK2) irradiated up to $3 \cdot 10^{15}$ n_{eq}/cm^2, where each line connects points belonging to the same sensor. The thickness of the active bulk varies in the range 45–55 µm. Several relevant features can be extracted from this plot:

1. High gain at low bias voltage does not yield good performances.
2. Increasing the bias voltage improves the temporal resolution almost in all cases, as the value of the shot noise remains smaller than the electronic noise floor even at the highest fluence.
3. Irrespective of the irradiation level, there is a broad area delimited by a bias in the range $150 - 650$ V and a gain > 15 where optimum performance is achieved. The irradiation levels are null (solid line), $\Phi = 8 \cdot 10^{14}$ n_{eq}/cm^2 (dashed), $\Phi = 1.5 \cdot 10^{15}$ n_{eq}/cm^2 (dashed - dot), and $\Phi = 2.5 \cdot 10^{15}$ n_{eq}/cm^2 (dashed-dot-dot).
4. The maximum bias is about 750 V. Above 700 V, it is rather common to produce irrecoverable damage to the sensor.

As explained in Section 2.9, the gain reduction caused by the acceptor removal mechanism can be mitigated with three techniques:

1. Add carbon to the gain implant to reduce the acceptor removal rate with fluence.

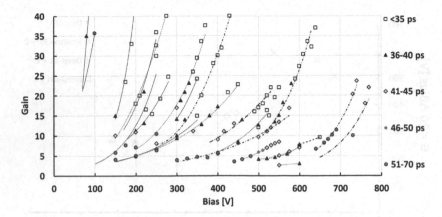

Figure 6.15 Temporal resolution in the gain-bias plane for a collection of 50 μm-thick UFSDs. The value of the resolution is indicated using different symbols for five ranges. This collection indicates that the best performances are obtained in the intervals of gain and bias given by gain > 15, 150 V <bias< 650 V. The irradiation levels are null (solid line), $\Phi = 8 \cdot 10^{14}$ n_{eq}/cm^2 (dashed), $\Phi = 1.5 \cdot 10^{15}$ n_{eq}/cm^2 (dashed-dot), and $\Phi = 2.5 \cdot 10^{15}$ n_{eq}/cm^2 (dashed-dot-dot).

2. Move the gain implant position deeper to increase the recovering capability of the bias voltage.
3. Reduce the lateral spread of the gain implant, as the impact of acceptor removal is lower at higher doping density.

The following plots analyze the effects of these three techniques. They display again the temporal resolutions in the gain-bias plane, using different symbols to indicate the value of the temporal resolution, divided in five ranges. The two plots of Figure 6.15 report the resolutions of two sensors with a shallow-B gain implant. The *top* panel shows the results for a sensor with a shallow-BH gain implant, while the *bottom* plot for shallow-BL + Carbon. In the *top* plot, the bias voltage needed to maintain gain = 10, Bias($G = 10$), moves by 500 V after a fluence $\Phi = 1 \cdot 10^{15}$ n_{eq}/cm^2, while in the *bottom* plot, the Bias(G = 15) point moves only by 370 V after the fluence $\Phi = 1.5 \cdot 10^{15}$ n_{eq}/cm^2. The combined effects of carbon co-implantation and low-temperature annealing reduce the voltage increase by about 200 V. Both sensors reach a temporal resolution less than 35 ps at fluences of 1–$1.5 \cdot 10^{15}$ n_{eq}/cm^2, and of 40–45 ps at 2.5–$3 \cdot 10^{15}$ n_{eq}/cm^2.

In Fig. 6.17, the temporal resolutions of two sensors with deep-BL gain implant, one without (*top*) and one with carbon (*bottom*) are shown. The sensors measured in the *bottom* plot have a gain layer that incorporates all three techniques to increase the radiation resistance: the gain implant is carbonated, deep, and narrow.

When new, both types of sensor work at relatively low bias, achieving a temporal resolution of about 40–50 ps. The presence of carbon is beneficial also for deep gain implants: after a fluence of $2.5 \cdot 10^{15}$ n_{eq}/cm^2, the bias($G = 20$) point changes

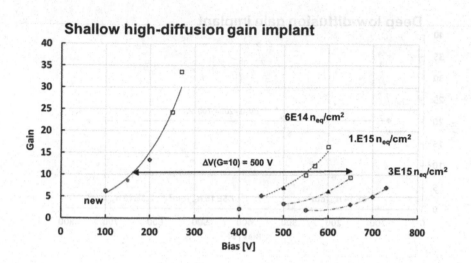

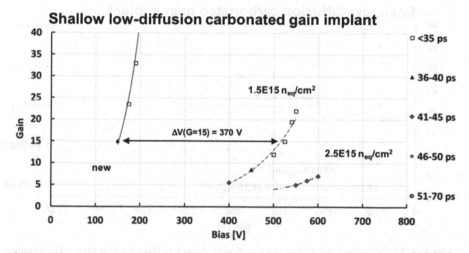

Figure 6.16 Temporal resolution as a function of fluence in the gain-bias plane. *Top*: shallow-BH gain layer. *Bottom*: shallow-BL carbonated gain layer (HPK ECX20840, UFSD3).

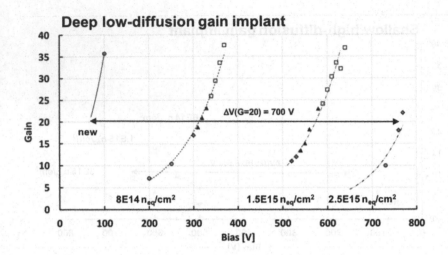

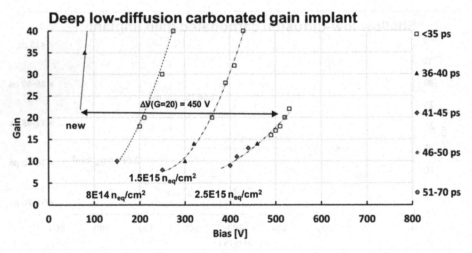

Figure 6.17 Temporal resolution as a function of fluence in the gain-bias plane. The value of the resolution is indicated using different symbols for five ranges. *Top*: deep-BL gain implant. *Bottom* : deep-BL carbonated gain implant (HPK2, UFSD3.2) [15].

by $\Delta V = 700$ V in the *top* plot while only by $\Delta V = 450$ V in the *bottom* plot. The deep-L carbonated gain layer is currently the most radiation-resistant design, and the only one able to maintain a temporal resolution below 35 ps even after being exposed to a fluence of $2.5 \cdot 10^{15}$ n_{eq}/cm^2. Another important benefit of this design is the possibility of obtaining equally good performance operating the sensors at lower bias voltages.

6.4.3   UFSDS NOISE

UFSDs owe their excellent performances to the presence of gain. As explained in Section 2.4, the shot noise increases proportionally to the value of gain multiplied by the excess noise factor, see Eq. (2.10). Given the relatively low value of gain and the small hole impact ionization coefficient, the noise contribution to the temporal resolution is negligible up to fluences around $2 \cdot 10^{15}$ n_{eq}/cm^2 [82]. Figure 6.18 reports the gain and noise at three values of fluence for a 45 μm-thick UFSD with a deep-BL gain implant (HPK2) [30]. The value of noise is almost constant up to the highest tested fluence, $\Phi = 2.5 \cdot 10^{15}$ n_{eq}/cm^2, a feature that allows reaching excellent temporal resolutions even for heavily irradiated sensors.

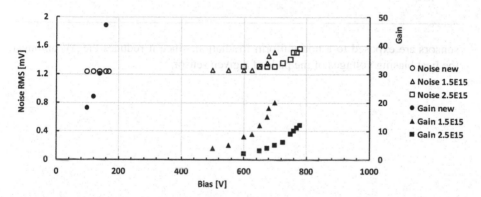

Figure 6.18   The noise (left *y*-axis) and gain (right *y*-axis) values of a new and two irradiated 45 μm-thick UFSD (HPK2) as a function of the bias voltage. The noise baseline RMS value, $\sigma = 1.25$ mV is due to the front-end electronics.

6.4.4   UFSDS NON-UNIFORM IRRADIATION

In experiments, quite often the area covered by the detectors is exposed to a non-uniform irradiation. For UFSDs, this fact leads to a gradient in gain: since the bias is common to the whole detector, the area exposed to lower radiation has a higher gain. For a given value of the bias voltage chosen to achieve the best performance for the high gain region, regions with lower gain are under-biased, and their performances might be degraded.

In the timing layer detector of the Phase2-CMS experiment, the dimensions of the UFSD sensors are foreseen to be 2×4 cm^2. Given the expected non-uniform irradiation map in CMS [9], the fluence can differ by as much as 20% over 4 cm. Under this condition, one side of the sensor would be exposed to a given fluence, for example, $\Phi = 1\cdot10^{15}$ n_{eq}/cm^2, while the other to a 20% larger fluence, $\Phi = 1.2\cdot10^{15}$ n_{eq}/cm^2. The signal reductions in various gain layer designs due to a 20% fluence difference are reported in Table 6.4. The table shows that, without carbon, the signal reduction exceeds 50% while, in carbonated gain implants, the reduction is about 35%. Therefore, the presence of carbon is beneficial also when the

Table 6.4

Collected Charge Loss Induced by a 20% Difference in Fluence, at $1\cdot10^{15}$ n_{eq}/cm^2 and Bias Providing 15 fC, for Different Gain Layer Designs

Gain Implant	Q [fC] @ $1\cdot10^{15}$ n_{eq}/cm^2	Q [fC] @ $1.2\cdot10^{15}$ n_{eq}/cm^2	Loss
Shallow+C	15	9.76	35%
Shallow	15	6.6	56%
Deep+C	15	9.7	35%
Deep	15	6.75	55%

sensors are exposed to a non-uniform irradiation, since it reduces the ΔV amongst the best biasing voltages of the pads of a given sensor.

7 Sensors for Extreme Fluences

The next generation of hadron colliders for particle physics will require tracking detectors able to efficiently record charged particles in harsh radiation environments, where expected fluences exceed 10^{17} n_{eq}/cm^2. Therefore, one of the most important goals of present R&Ds on silicon sensors is to increase their radiation tolerance by more than an order of magnitude [83, 1].

This chapter opens by outlining the present understanding of radiation damage at high fluences. In the second part, a possible approach to the design of sensors capable of operating at extreme fluences is discussed, addressing different aspects such as the choice of active substrate thickness, the gain layer design, the optimization of the sensor edge and inter-pad regions. Finally, the ongoing experimental work to extend the simulation models in the fluence regime above 10^{16} n_{eq}/cm^2 is discussed. Additional material on these topics can be found in [78, 127, 77].

7.1 THE REGIME OF EXTREME FLUENCES

As described in Section 1.3, irradiation causes 4 main macroscopic effects: (i) the increase of the dark current due to the creation of electron-hole generation centres, (ii) changes in doping density, leading to the increase of the bias voltage necessary to fully deplete the sensor (proportional to the effective acceptor density of the substrate), (iii) the reduction of charge collection efficiency due to trapping of the charge carriers, and (iv) a decrease of the charge carrier mobility. As an example, due to these effects, a 150 μm-thick n-in-p silicon sensor can efficiently operate only up to $\Phi = 1$–$2 \cdot 10^{16}$ n_{eq}/cm^2 [71]. At higher fluences, only a fraction of the sensor thickness can be depleted, reducing considerably the signal amplitude and making it impossible to efficiently detect particles.

The 3D silicon sensor architecture, with n- and p-electrodes etched inside the silicon substrate [107], has a superior tolerance to radiation; it has been demonstrated that 3D sensors function correctly up to $\Phi = 2.5 \cdot 10^{16}$ n_{eq}/cm^2 [44]. The strength of this design is the decoupling between the active sensor thickness from the lengths of the electrons and holes collection paths, allowing to have thick sensors with a short collection path. However, when irradiated at a fluence of $\Phi \sim 10^{17}$ n_{eq}/cm^2, 3D sensors show a collected charge of only about 0.5 fC [56] due to the decrease with irradiation of the charge collection efficiency. The CCE might be recovered by reducing the distance among the electrodes, at the cost of increasing the difficulties related to the production of sensors with a very high density of electrodes.

In the past few years, the study of the properties of heavily irradiated silicon sensors has demonstrated that they behave better than expected: above

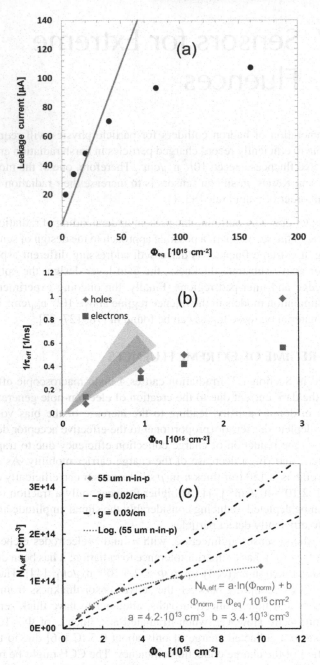

Figure 7.1 (*a*) Leakage current as a function of fluence (dots), compared to the expected full bulk generation current using a damage constant $\alpha(-23\ °C) = 3.48 \cdot 10^{19}$ cm^{-2} (line) [34]. (*b*) The effective trapping probabilities for electrons (square) and holes (diamonds) as a function of the fluence, compared to the predictions with $\beta_e = 3.5 \pm 0.6$ cm^2/ns and $\beta_h = 4.7 \pm 1.0$ cm^2/ns, respectively (bands). [96]. (*c*) Bulk effective doping concentration as a function of fluence (diamonds) with a logarithmic fit superposed. Predictions from the linear model with two different values of $g_{eff} = 0.02$ and 0.03 cm^{-1} are also shown in [49].

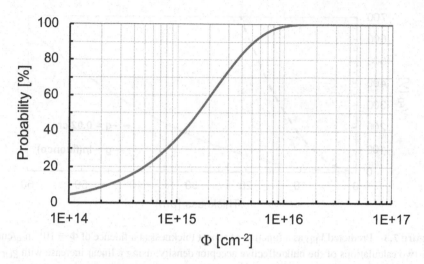

Figure 7.2 Probability that a circular surface with $r = 1.18 \cdot 10^{-8}$ cm is crossed by a particle, as a function of the fluence.

$\Phi = 5 \cdot 10^{15}$ n_{eq}/cm^2, the effects induced by the irradiation tend to saturate, evolving from linear to logarithmic as a function of fluence.

Figure 7.1 illustrates the saturation effect for three quantities. The *top* plot shows the evolution of the leakage current measured in irradiated p^+-n 'spaghetti' diodes [34] (dots), compared to the predicted bulk generation current (see Eq. (1.8)) using a damage constant $\alpha(-23\ °C) = 3.48 \cdot 10^{19}$ cm^{-2} (line). The *middle* plot reports the measured effective trapping probabilities for electrons (square) and holes (diamonds) as a function of the fluence, compared to the predictions using a linear extrapolation model. The effective trapping damage constants used are $\beta_e = 3.5 \pm 0.6$ cm^2/ns and $\beta_h = 4.7 \pm 1.0$ cm^2/ns, respectively (bands) [96]. The last plot shows the bulk effective doping concentration as a function of fluence (diamonds), with a logarithmic fit superposed, compared to the predictions (see Eq. (1.9)) using two different values of g_{eff}, 0.02 and 0.03 cm^{-1} [49].

The saturation of radiation effects observed in highly irradiated sensors is still under investigation and its origin is not yet understood. From the microscopic point of view, a possible explanation is that at high fluence defects are created in an already damaged lattice, and the properties of the silicon substrate are not further modified. Considering that the distance between two atoms inside the reticle is $r_{Si} = 1.18 \cdot 10^{-8}$ cm, the probability that every lattice cell has been crossed by a particle becomes 1 at $\Phi = \sim 10^{16}$ cm^{-2}; for higher fluences, particles are crossing an area that on average has already been crossed at least once (see Fig. 7.2).

The saturation of radiation damages represents a key element in the design of sensors able to sustain extreme fluences, as it opens the way to some viable solutions. The bias needed to fully deplete a silicon sensor V_{FD}, Eq. (1.1), is proportional to the

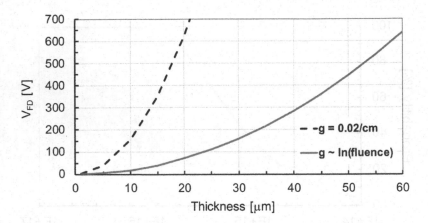

Figure 7.3 Predicted V_{FD} as a function of sensor thickness at a fluence of $\Phi = 10^{17}$ n_{eq}/cm^2, for two calculations of the bulk effective acceptor density: using a linear increase with $g_{eff} = 0.02$ cm^{-1} (dashed line), or using g_{eff} extrapolated from the data (line) [49].

effective acceptor density $N_{A,eff}$ and the square of the sensor active thickness d :

$$V_{FD} \propto N_{A,eff}d^2. \tag{7.1}$$

Considering, as an example, an irradiated 100 μm thick sensor, once the effective acceptor density of the bulk reaches values of the order of $2 \cdot 10^{14}$ atoms/cm^3, a bias voltage higher than 1000 V is needed to fully deplete it.

The depletion voltage as a function of thickness at $\Phi = 10^{17}$ n_{eq}/cm^2 is shown in Fig. 7.3, assuming a linear (dashed line) or logarithmic (solid line) dependence of $N_{A,eff}$ on the fluence. Thanks to the saturation of the creation rate of the acceptor-like states in the bulk, shown in Fig. 7.1 *bottom*, thin substrates can be still depleted with a relatively low bias: in particular, sensors thinner than 50 μm are fully depleted with less than 500 V. In general, only very thin sensors can be fully depleted without getting close to their breakdown limit, in the 10^{17} n_{eq}/cm^2 fluence regime.

Thin substrates yield an obvious problem: the signal generated by an impinging MIP is rather small. The most probable value of the collected charge for a MIP crossing a 50 μm thick sensor is 0.5 fC (about 0.1 fC each 10 μm). The most radiation-resistant front-end electronics, currently under development for the High-Luminosity LHC tracking systems, requires at least 1 fC of charge from the sensor to efficiently record a particle signal [10]. Hence, to generate a pulse visible in the electronics, thin sensors need internal multiplication of signal: a 20 μm-thick sensor needs for example an internal gain of at least 5.

As described in Section 2.1.1, internal gain due to impact ionization occurs when the electric field inside the sensor reaches the critical value of $\mathscr{E}_C \sim 25$ V/μm. The main requirements to have stable signal multiplication can be summarized as follows: (i) the electric field should be near the critical value for a short distance, of

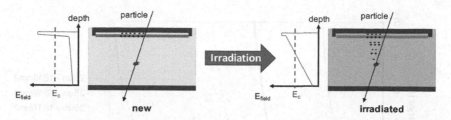

Figure 7.4 Schematic cross-cut of a new (*left*) and irradiated (*right*) UFSD, and their respective electric field profiles as a function of the sensor depth. Different tones of grey reflect the doping concentrations in the sensor. The detail of the creation and drifting of an electron-hole pair is shown as an example.

the order of 1 μm, (ii) the field above the critical value should be as flat as possible to avoid uncontrolled multiplication, and (iii) the field value should be controlled by bias and not by bulk doping. These requirements are well satisfied by the UFSD design, yielding a moderate internal gain and stable operation. Under irradiation, the effective acceptor density of the sensor changes according to Eq. (1.9): the highly doped gain implant becomes less and less doped, whereas the effective bulk doping increases. As a consequence, the signal multiplication occurring in the gain implant progressively decreases with irradiation, while, as the bias voltage increases to maintain the sensor fully depleted, the field in the bulk becomes high enough to generate gain, as depicted in Fig. 7.4. Irradiation decreases the mean free path of charge carriers, so, even at high electric fields, the charge multiplication is quenched, and it is possible to obtain a moderate gain, controlled by the applied bias. Multiplication in the bulk will be present up to fluences at which the mean free path of electrons and holes is longer than $\alpha(\mathscr{E})^{-1}$, the inverse of the ionization coefficient described in Eq. (2.1). It is crucial that the sensor can be over-depleted even at the highest fluences, so that the electric field remains as flat as possible when it approaches the critical value and does not lead to an uncontrolled breakdown.

A range of sensor thickness that might represent a working compromise is 20–35 μm, as sensors can still be depleted after a fluence of $\Phi = 10^{17}$ n_{eq}/cm^2, and a relatively low gain of 5–10 is sufficient to guarantee the delivery of at least 1 fC of charge, as shown in Fig. 7.5.

If, in addition to position, the timing information is also required, then the requests on the sensor performances become more demanding. Good timing capability, for example, a temporal resolution of about 50 ps, is achieved if the sensor provides at least 5–6 fC of charge to the electronics [73]. For a 20–35 μm-thick UFSD sensor, this means achieving an internal gain of ~ 20 up to the target fluence. Such high gain is easily reached using a gain layer in not-irradiated sensors, while it is a much harder goal in heavily irradiated sensors when the gain is generated only in the sensor bulk. As discussed in Section 6.2, the addition of carbon atoms in the volume of the gain implant more than doubles its radiation tolerance, extending the fluence range in which the device is capable of delivering the necessary charge. The increase

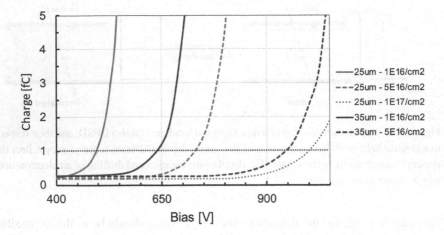

Figure 7.5 Simulated collected charge for 25 μm- and 35 μm-thick UFSD sensors irradiated at $\Phi = 1\text{--}10 \cdot 10^{16}$ n_{eq}/cm^2, as a function of the applied bias voltage. The simulation is performed with the WF2 program; the Massey model [100] is used for the gain mechanism.

of radiation resistance of the gain implant by an order of magnitude requires still further R&D work.

7.2 SENSOR DESIGN

Figure 7.6 shows how the bulk (solid lines) and gain implant doping densities (dashed lines) (described in Fig. 2.24) evolve with fluence, with or without taking into account the two main effects introduced in Section 6: (i) the saturation of the effective acceptor creation in the bulk, and (ii) the reduction of the acceptor removal mechanism due to carbon co-implantation in the gain implant volume. Both effects help preserving the internal charge multiplication to higher fluence values. To reach $\Phi = 10^{17}$ n_{eq}/cm^2, further developments of both bulk substrate and gain layer design have to be pursued. The current line of studies includes (i) substrate optimization in terms of bulk doping and thickness, (ii) electric field engineering via gain layer design, and (iii) defect engineering both in the sensor substrate and in the gain implant volume. In addition, the sensor periphery and pad isolation designs need to be optimized to avoid edge or inter-pad breakdown at very high fluences, when the bias voltage applied creates an electric field ≥ 15 V/μm.

7.2.1 SUBSTRATE CHOICE

The present knowledge on substrate defect engineering relies on many years of R&D aimed at improving the radiation hardness of silicon sensors. Oxygen-enriched substrates are considered more resilient to acceptor creation, as oxygen interstitials, O_i, inhibit the generation of deep acceptor states, while oxygen dimers, O_{2i}, participate

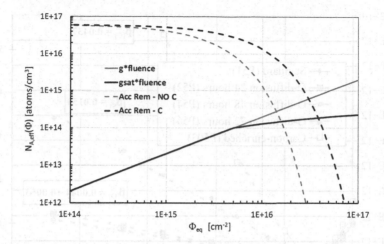

Figure 7.6 Solid lines: evolution of the effective acceptor density in the bulk using either $g_{eff} = 0.02$ cm^{-1} or the saturated g_{eff} value. Dashed lines: evolution of the effective acceptors density in the gain implant with or without carbon co-implantation.

in the generation of shallow donor states in the material [42]. The presence of oxygen has a direct impact on the sensor depletion voltage, as shown in Fig. 7.7: for oxygen-enriched substrates, full depletion voltage exhibits a milder increase with fluence. However, the effect of high oxygen density on the gain implant preservation has not been studied yet. It has been suggested that high oxygen densities might fasten the acceptor removal mechanism [103].

An important aspect in the design of thin sensors for extreme fluence is the control of the impurities introduced in the active bulk by the detector handle wafer. Since the active bulk of the sensor is very thin, impurities present in the low resistivity support might diffuse during processing into the whole sensor active thickness. Two main types of thin sensors have been explored so far: silicon-on-silicon (Si-on-Si), where a float zone high resistivity silicon wafer is bonded to a low resistivity Czochralski (Cz) wafer that serves as a support, and epitaxially grown substrates on low resistivity Cz supports. The differences between the two processes are the cost and the contamination of the active sensor volume. Czochralski supports have a high oxygen concentration, of the order of 10^{18} cm^{-3}. In the Si-on-Si process, the migration of impurities from the handle wafers into the active volume hardly occurs due to direct bonding. On the other hand, in epitaxial silicon grown on Cz substrate this migration happens, and the oxygen concentration exceeds $5 \cdot 10^{17}$ cm^{-3} in the 30 μm closest to the contact surface [42].

Another interesting parameter to be investigated is the silicon crystal orientation: at present, $< 100 >$ silicon substrates are commonly used. However, $< 110 >$ and $< 111 >$ can be studied to understand if different lattice orientations have a beneficial effect on the mobility of the carriers at high fluences.

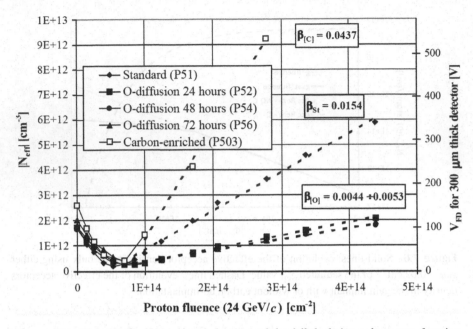

Figure 7.7 The effective space charge density and the full depletion voltage as a function of the proton fluence for standard silicon sensors, carbon-enriched, and three types of oxygen diffused samples: 24, 48, and 72-hour diffusion at 1150 °C [11].

7.2.2 THE GAIN LAYER REGION

The gain layer contributes to the sensor gain up to fluences of $5-10 \cdot 10^{15}$ n_{eq}/cm^2. For higher fluences, the acceptor removal mechanism reduces the active acceptors density of the gain implant to a value where its contribution to the electric field is too low to trigger impact ionization. The most radiation resistant LGADs currently produced suggest that the gain implant doping can still play a role in the signal multiplication above $\Phi = 10^{16}$ n_{eq}/cm^2. How much this contribution is and how to extend it as much as possible is presently under study.

As discussed in Section 2.1.1, the charge multiplication factor of a gain layer depends on the electric field $\mathscr{E}$ and the length of the high-field region (the distance from the gain implant to the n^{++} electrode). As a consequence, a given gain value might be obtained with different designs of the gain implant doping profile. A shallow boron profile, closer to the n^{++} junction, leads to a highly-peaked and narrow high-field region; while deeper, less doped boron profiles generate wider and lower electric fields. These two designs lead to different residual gain capabilities after irradiation: a deep gain layer implant provides faster recovery of the gain with bias, while a shallow implant has a higher residual acceptor density (since it starts from a higher initial density).

The addition of carbon in the gain implant volume improves the radiation resistance by a factor of about 3 (e.g., see Fig. 6.5). Additional developments can further exploit defect engineering. As for the case of carbon co-implantation, such engineering aims to reduce boron removal or compensate for it with new acceptor-like states induced by irradiation. Impurities such as gold, tin, or transition metals can introduce such acceptor levels in the silicon band-gap. However, their implantation is more difficult as it needs to be done at high energy (given their atomic mass), and it might lead to strong radiation damage into the lattice. Another interesting element is germanium. According to studies aiming at increasing the lifetime of solar panel [80], germanium might enhance the gain implant tolerance to radiation. In concentration above 10^{19} cm^{-3}, germanium reduces the creation of B-O complexes, which are the main culprit for the acceptor removal mechanism. In fact, it has been measured that germanium inhibits the diffusion of O_{2i} defects in the silicon lattice [79], protecting boron atoms in the reticle. Therefore, to further extend the gain implant survival with radiation, it can be interesting to investigate the effect of germanium co-implantation in the gain implant volume. Unfortunately, given the high atomic mass, germanium requires high implantation energy, steeply increasing the production costs, and can cause distortion in the silicon crystal structure due to the bigger size of the germanium atoms.

7.2.3   THE SENSORS PERIPHERY

According to the predictions shown in Fig. 7.5, a reverse bias in excess of 800 V has to be applied to extremely irradiated ~ 25 μm-thick sensors in order to reach the target charge collection. The guard-ring structure, introduced in Section 2.8 and sketched in Fig. 2.21, needs to be optimized for small thicknesses as they need to sustain very high bias values over distances comparable to the sensor thickness. At high fluences, the concurrent action of bulk and surface damage has to be taken into account in the design: on the one side, the effect of fixed charges introduced in the oxide region between rings tend to short together the n^{++} implants, while, on the other side, the trap defects created in the bulk counterbalance this effect. How the interplay of these two effects evolve up to about $\Phi = 10^{17}$ n_{eq}/cm^2 will be studied in future productions and irradiation campaigns.

The typical guard-ring structure of n-in-p sensors needs to be revised and optimized for very thin substrates. The p-stop between guard-rings float to a potential between the bias voltage and ground: given the very reduced thickness of the bulk, the p-stop might float to potential values quite close to the bias value. Under this condition, the sequence of guard-rings might not be able to sustain a large voltage drop. Simulations also show the presence of very high electric field peaks near the p-stop that can trigger a breakdown. An interesting approach, currently under consideration, is a design of the guard-rings without the p-stop isolation. In this approach, before irradiation, the guard-rings are shorted together by the n^{++} inversion layer. With increasing irradiation, the bulk damage provides isolation, and the guard-rings float at different potential values, allowing a smooth transition from ground to the bias potential at the edge.

Finally, also the distance between the last guard-ring and the highly p-doped cut line needs to be tailored to the design of thin sensors. It is likely that for small thicknesses, the rule of thumb [116] of using a distance of about three times the substrate thickness between the cutting edge and the sensitive region will be superseded.

7.2.4 THE INTER-PAD REGION

The inter-pad structures described in Section 2.8 (see Fig. 2.22) and extensively measured and characterized in Section 5.2, need to be optimized to work up to $\Phi = 10^{17}$ n_{eq}/cm^2, avoiding cross talk between pads or early breakdown of the sensor. Inter-pad structures with JTE and p-stop will likely ensure good pad isolation up to extreme fluences. However, the distance between the gain implant, the n-deep, and p-stop, as well as the field plate extension, needs to be optimized to avoid high field regions, which at high fluences can trigger a premature breakdown. The performance evolution with irradiation of more aggressive inter-pad strategies, such as TI-LGADs or RSD, is presently under study.

7.3 SIMULATION OF EXTREMELY IRRADIATED SENSORS

To successfully design and operate thin silicon sensors at a fluence of $\Phi=10^{17}$ n_{eq}/cm^2 and above, a close interconnection between experimental activity and simulation is crucial. Models are needed to identify the optimal design able to sustain extreme fluences, including dedicated studies on the optimal gain layer design. As an example, in Fig. 7.9 it is possible to see the leakage current behavior as a function of the fluence for a 25 μm-thick LGAD with the gain implant designs introduced in Fig. 7.8. To account for the acceptor removal mechanism due to radiation, the state-of-the-art parametrization ($k_{cap}N_{int} = 0.23$) of the c factor as a function of the peak doping concentration is used, and the bulk and surface damages follow the 'Perugia 2020' updated model [16, 29]. The impact ionization mechanism is calculated via the Massey model [100]. The increase of the leakage current reflects the internal multiplication of the charge carriers induced by impact ionization: when new, sensors have a similar gain behavior. At a fluence of $5 \cdot 10^{15}$ n_{eq}/cm^2, the residual gain implant of the shallow profile still contributes to the local increase of the electric field, triggering signal multiplication about 50 V earlier than with the deep-profile. However, it has to be verified experimentally whether the electric field value is too high for stable operation, causing a sizable noise increase. For a fluence of $\Phi = 10^{16}$ n_{eq}/cm^2, the advantages of having a shallow gain implant profile appear reduced.

Experimental results are necessary to extend the present models to fluences above $\Phi = 10^{16}$ n_{eq}/cm^2. Figure 7.10 shows the CCE from 55 μm-thick n-in-p sensors irradiated up to $\Phi = 10^{16}$ n_{eq}/cm^2, compared with predictions from the WF2 program. The WF2 CCE predictions, extrapolated up to $\Phi = 5 \cdot 10^{17}$ n_{eq}/cm^2, show an encouraging forecast of the CCE from thin sensors. Furthermore, recent results on extremely irradiated silicon sensors [41, 132] showed that above $\Phi = 10^{16}$ n_{eq}/cm^2 the charge collection efficiency is higher than expected. Therefore, an update of the existing models up to extreme fluences is necessary. The extension of the predictive capability of the already developed radiation damage models to irradiation levels

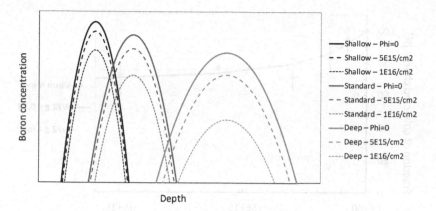

Figure 7.8 Gaussian gain implants at different depths from the sensor surface. The gain implant profiles before irradiation (solid lines) are compared with their evolution after $\Phi = 5 \cdot 10^{15}$ n_{eq}/cm^2 (dashed lines) and $\Phi = 10^{16}$ n_{eq}/cm^2 (dotted lines). The initial acceptor removal is calculated via Eq. (6.1), with $k_{cap}N_{Int} = 0.23$.

above $\Phi = 10^{17}$ n_{eq}/cm^2 is not straightforward: some of the radiation damage effects, usually negligible at lower fluences, have to be taken into consideration. These

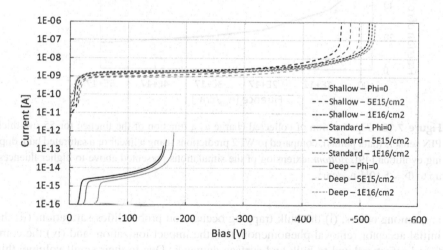

Figure 7.9 Simulated I(V) characteristics for a 25 μm-thick sensor with bulk doping $1 \cdot 10^{13}$ atoms/cm^3 and gain layer designs as from Fig. 7.8. The leakage current evolution with bias is shown before irradiation (solid lines), after $\Phi = 5 \cdot 10^{15}$ n_{eq}/cm^2 (dashed lines), and $\Phi = 10^{16}$ n_{eq}/cm^2 (dotted lines). The 'Perugia 2020' updated model has been used to account for radiation-induced bulk and surface damage [16, 29]. Simulation from [130] using TCAD Synopsys [128].

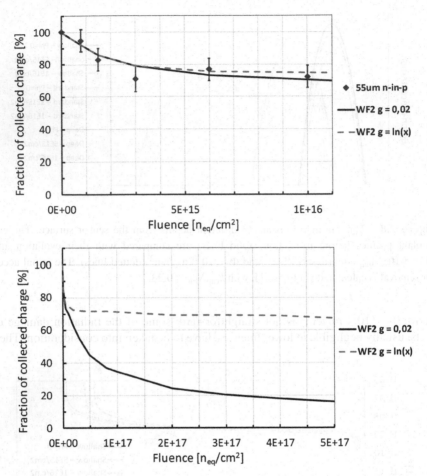

Figure 7.10 *Top*: fraction of collected charge as a function of the fluence for 55 μm-thick PIN sensors (data points), compared to WF2 predictions using a linear or a saturated bulk doping evolution [49]. *Bottom*: extension of the simulations presented above to higher fluences, up to $\Phi = 5\cdot10^{17}$ n$_{eq}$/cm^2.

are, among others, (i) the bulk trapping occupation probabilities saturation, (ii) the initial acceptor removal phenomenon, (iii) the impact ionization, and (iv) the combined effects related to bulk and surface damages. Due to their small volume, thin silicon sensors represent the ideal framework to study the radiation effects in a new fluence regime and to extend existing models to fluences above 10^{16} n$_{eq}$/cm^2.

A Productions

The studies presented in this book are based on sensors produced by three vendors, Centro Nacional de Microelectronica (CNM) in Barcelona (ES) [6], Hamamatsu Photonics (HPK) in Hamamatsu (JP) [109], and Fondazione Bruno Kessler (FBK) in Trento (IT) [94]. The first results of LGADs properties were presented by CNM in 2014 [40], CNM produced the first 50 μm-thick UFSDs in 2016. The second vendor to produce UFSDs was HPK; the first results on HPK-UFSDs have been shown at the TREDI 2017 conference [47]. At the end of 2018, HPK produced the first prototypes for the CMS and ATLAS MIP Timing Detector R&D activities. FBK manufactured its first production of LGADs (300 μm-thick) in 2016 [38], followed by several 50 μm-thick UFSDs productions: UFSD2 in 2017, UFSD3 in 2018, UFSD3.1, RSD1 and TI-LGAD in 2019, and UFSD3.2 in 2020.

The main features of the UFSD productions are detailed below.

A.1 FONDAZIONE BRUNO KESSLER

A.1.1 UFSD1

FBK, in 2016, completed its first production of LGAD sensors, called UFSD1. These sensors were manufactured using Si-on-Si 6" p-type Float Zone (FZ) wafers, with a ~ 275 μm-thick high resistivity ($\rho > 5$ kΩcm) bulk.[1] The main goal of this production was to investigate different doses of the gain implant and calibrate the FBK processing steps. Twelve wafers, with five different gain doses of boron, spaced by 2%, have been produced. UFSD1 includes a large variety of structures: strip and pixel sensors with different pitches, array, and single pad sensors with pad size in the range of 0.25–25 mm^2. This production is double-sided and it includes devices with patterned electrodes on the p-side [38], also known as *inverted*-LGAD.

A.1.2 UFSD2

The first production of thin LGADs by FBK, called UFSD2, was completed in 2017. This production consists of 18 Si-on-Si 6" p-type float zone wafers, with a ~ 60 μm-thick high-resistivity ($\rho > 3$ kΩcm) bulk, thermally bonded on a support wafer 500 μm-thick. The thermal bonding reduced the active wafer thickness from the nominal 60 μm value to ~ 55 μm.

The targets of this production were to establish a reliable design of UFSDs and test solutions to improve the gain implant's radiation resistance. For these reasons, different gain implants have been implemented: the acceptor atoms of the gain implant are boron and gallium. For the first time, the gain implant of several wafers

---

[1] The term high-resistivity indicates bulk with a resistivity ρ in a range 1–10 kΩcm.

has been enriched with carbon. Carbon has been infused only in the gain implant volume to avoid a sharp increase in the leakage current. Five shallow-gain implant designs have been implemented in this production: (i) Boron High Diffusion (BH), (ii) Boron Low Diffusion (BL), (iii) Gallium (Ga), (iv) Carbonated Boron High Diffusion (BH + C), and (v) Carbonated Gallium (Ga + C). The high and low diffusion labels refer to the thermal load used to activate the gain implant. All Ga-gain implants have been processed with a low thermal load due to the greater diffusivity of gallium. Ten wafers have a boron-doped gain implant, with four dose splits of increasing doping spaced by 2%. The eight remaining wafers are gallium-doped with three dose splits, spaced by 4%. Dose 1.00 a.u. represents the p-reference dose.

Four boron- and two gallium-gain implants have been enriched with two different carbon doses: Low (1 a.u., reference dose) and High (10 a.u.). All information on dopants, gain layer doses, carbon doses, and diffusion process are reported in Table A.1.

Table A.1
Wafers of the UFSD2 Production

Wafer #	Dopant type	p-gain dose^a	Carbon dose^a	Thermal load	V_{GL} [V]
1		0.98	-	Low	23.1 ± 0.5
2		1.00	-	Low	23.9 ± 0.5
3		1.00	-	High	22.8 ± 0.5
4		1.00	1	High	22.5 ± 0.5
5		1.00	10	High	4.0 ± 0.5
6	Boron	1.02	1	High	23.0 ± 0.5
7		1.02	10	High	-
8		1.02	-	High	-
9		1.02	-	High	-
10		1.04	-	High	23.8 ± 0.5
11		1.00	-	Low	31.0 ± 0.5
12		1.00	-	Low	-
13		1.04	-	Low	-
14	Gallium	1.04	-	Low	31.5 ± 0.5
15		1.04	1	Low	30.5 ± 0.5
16		1.04	10	Low	11.0 ± 0.5
18		1.08	-	Low	-
19		1.08	-	Low	33.5 ± 0.5

^{a} In arbitrary unit.

In this production, wafers were manufactured using mask aligners, which pattern the entire wafer at once, see Fig. A.1. The UFSD2 wafer layout includes devices with a large variety of geometries and dimensions.

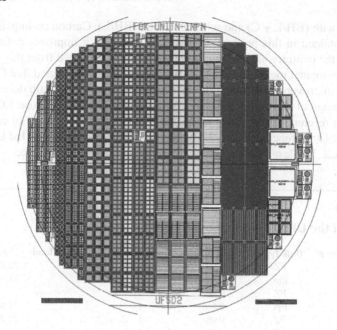

Figure A.1 UFSD2 wafer layout.

1. Single pad sensors with pad size of 1 mm^2, 4 mm^2, and 9 mm^2, as UFSDs technology demonstrators [76].
2. Pad arrays 2 × 2, 2 × 4, and 2 × 8, as precursors of segmented and large area sensors.
3. Strip sensors, suitable for beam counting applications in medical physics [18, 69, 20, 19].
4. Strip and pad sensors for soft X-ray (energy in a range 1–10 keV) detection [45].
5. Fat strips, developed for time measurements in the TOTEM and CT-PPS experiments [46, 106] at CERN.
6. Pixels matrix of 45 × 40 pads, each 300 × 300 μm^2, as the first demonstrator of UFSD pixel sensors usable with the bump bonding layout of the NA62 readout chip.

A.1.3 UFSD3

The FBK-UFSD3 production has been designed to investigate specific sensor features required by the CMS and ATLAS timing detectors: radiation hardness, a narrow inactive area between pads, and uniformity of segmented large area sensors. This production consists of 20 6″ wafers ~ 55 μm-thick with a high-resistivity bulk: 16 wafers have a float zone bulk while four wafers have an Epitaxial (Epi) bulk.

Four gain layer designs have been implemented in UFSD3: Boron High and Low

Diffusion with (BH/L + C) and without carbon (BH/L). Carbon co-implantation has been maintained in this production as UFSD2 showed it improves radiation resistance. On the contrary, the Ga-gain implant has been excluded from this production since experimental measurements (see Section 6.2.2) demonstrated that Ga does not lead to an improvement in radiation resistance. Five splits of p-gain dose in steps of 2%, in a range of dose 0.96–1.04 a.u., have been implemented. Dose 1.00 a.u. represents the reference dose, equal to that used in UFSD2. Four doses of carbon have been used: 1 (reference dose), 2, 3, and 5 a.u. The UFSD3 full wafer list is presented in Table A.2. The last column reports the gain layer depletion voltage.

Table A.2
Wafers of the UFSD3 Production

Wafer #	Bulk type	p-gain dose[a]	Carbon dose[a]	Thermal load	V_{GL} [V]
1	FZ	0.98	-		23.0 ± 0.3
2	Epi	0.96	-		21.0 ± 0.3
3	FZ	0.96	1		-
4	Epi	0.96	1		21.0 ± 0.3
5	FZ	0.98	1		22.3 ± 0.3
6	FZ	0.96	2	Low	-
7	FZ	0.98	2		19.0 ± 0.3
8	FZ	0.98	2		-
9	FZ	0.98	3		15.8 ± 0.2
10	FZ	1.00	3		-
11	FZ	1.00	5		11.6 ± 0.2
12	FZ	1.02	-		22.9 ± 0.3
13	Epi	1.00	-		22.6 ± 0.3
14	FZ	1.02	1		23.4 ± 0.3
15	Epi	1.00	1		22.0 ± 0.3
16	FZ	1.02	2	High	-
17	FZ	1.02	2		-
18	FZ	1.04	2		20.3 ± 0.3
19	FZ	1.02	3		-
20	FZ	1.04	3		17.3 ± 0.2

[a] In arbitrary unit.

Contrary to UFSD2, the stepper lithographic technique was used in UFSD3. The stepper technology uses an area of 25×19 mm^2, called reticle, which is repeated several times on the wafer surface. Two stepper reticles have been used in UFSD3, Fig. A.2: reticle A contains a 4×24 pads sensor, each pad 1×3 mm^2 and a few smaller structures while reticle B has various geometries. Compared to the mask aligner technology, the strength of the stepper is the greater spatial precision, which reduces the distance between implants. Sensors larger than the reticle are produced, combining images from different exposures (photo-composition). To assess this technique validity, small sensor structures were produced in UFSD3 in a single shot and photo composed, finding no differences in their properties.

Reticle A   **Reticle B**

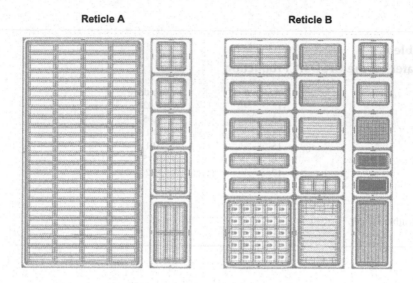

Figure A.2 The two reticles of the UFSD3 production.

In UFSD3, several new strategies for the gain layer termination implants (JTE and p-stop) have been pursued, aiming to minimize the no-gain region between pads. Four different layouts of the area between adjacent pads have been implemented: Aggressive, Intermediate, Safe, and Super Safe, with a nominal distance between gain implants of ~ 10 μm, ~ 20 μm, ~ 30 μm, and ~ 40 μm, respectively.

A.1.4 UFSD3.1

In 2019, UFSD3.1 was produced to investigate the effect of different p-stop implant designs and doses on the UFSDs capability to reach and hold high bias voltages, Section 5.2.3. UFSD3.1 consists of 7 6″ epitaxial wafers with a high-resistivity bulk 55 μm-thick. All wafers have a shallow-BH gain layer with the same p-gain dose. The wafers differ in the p-stop dose: six different doses in a range of 0.02–1 a.u. have been implanted (see Table A.3). Dose 1 a.u. represents the reference dose, equal to that used in the UFSD2 and UFSD3 productions.

UFSD3.1, as UFSD3, has been produced using the stepper lithographic technology, Fig. A.3 shows the reticle used in this production. The reticle layout includes 11 2×2 array devices, differing in the inter-pad design. The inter-pad layouts implemented in UFSD3.1 can be grouped in to four main categories: single p-stop, single p-stop + extra ring, double p-stop, and double p-stop + bias grid. Within each category, variations of the main layout are implemented. Table A.4 reports the 11

Table A.3
Wafers of the UFSD3.1 Production

Wafer #	p-stop dose[a]
12	0.02
13	0.05
14	0.1
15	0.1
16	0.15
17	0.2
18	1

[a] In arbitrary unit.

Table A.4
UFSD3.1 Inter-pad Designs

Inter-pad design	Inter-pad type	Inter-pad nominal distance [μm]
Single p-stop + extra ring	1	16
	6	28
	7	28
	8	28
Single p-stop	2	21
	3	21
	11	21
	4	24
	5	25
Double p-stop	9	38
Double p-stop + bias grid	10	49

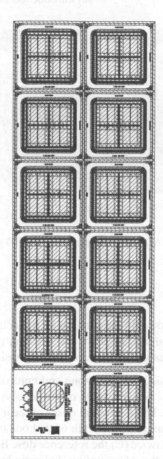

Figure A.3 UFSD3.1 reticle.

different inter-pad designs with the respective nominal distances between adjacent gain implants.

A.1.5 TRENCH-ISOLATED-LGAD

The TI-LGAD production of 2019 aims at demonstrating the feasibility of implementing thin LGADs segmented using the trench isolation technology. This is the first LGAD production using this technique.

The wafer layout consists of pairs of pads (Fig. A.4) with pad size $375 \times 250\ \mu m^2$, fabricated on 55 μm-thick p-epitaxial bulk. The layout splits implemented about 30 pairs, differing in the number of trenches (1 or 2), the dimension of the borders, and the guard-rings isolation method (p-stop or trench).

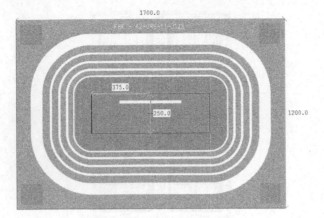

Figure A.4 Layout of an LGAD-LGAD pair in the TI-LGAD production (dimensions shown in μm).

A.1.6 RSD1

RSD1, in 2019, was the first FBK production of AC-coupled LGAD silicon sensor. The RSD1 production goal was to prove the working principle of the AC-LGAD design and find the best design parameters. RSD1 consists of 15 6″ p-type wafers, with 55 μm-thick high-resistivity float zone and epitaxial bulks ($\rho > 3$ kΩcm and $\rho > 1$ kΩcm, respectively). Several splits of the process have been implemented in this production, Table A.5. Three different doses of the n^{++} cathode have been used: a reference dose (B), approximately a tenth of that one used in standard UFSD productions, one half (A), and twice (C) the reference dose B. Two different thicknesses, Low (L) and High (H), of the AC-coupling dielectric have been implemented. The p-gain doses of 0.92, 0.94, and 0.96 a.u., normalized to the reference one, have been implemented. Finally, three p-doses (A, B, and C) were used in the p-stop terminations.

The RSD1 layout includes several device geometries, with different pitch and pixel size. The RSD1 reticle, shown in Fig. A.5, is divided into three parts: on the left side, there are matrices with 2×2, 3×3, 5×5, 8×8, and 10×10 pads, with different pad sizes and pitch from 50 to 300 μm. In the centre, there are larger matrices with 3×3, 4×4, and 5×5 pads, and on the right side, there are a 180 μm-pitch strip sensor and 64×64 square matrices (50 μm-pitch) [10].

A.1.7 UFSD3.2

The UFSD3.2 production, in 2020, was focused on the optimization of the gain layer and inter-pad designs to maximize the UFSDs radiation resistance. The production consists of 19 6″ p-type epitaxial wafers, with a high-resistivity bulk and active thicknesses of ~ 55 μm (3 wafers), ~ 45 μm (14 wafers), ~ 35 μm (1 wafer) and ~ 25 μm (1 wafer).

Table A.5
Wafers of the RSD1 Production

Wafer #	Bulk type	n-plus dose[a]	p-gain dose[a]	Dielectric thickness[a]	p-stop dose[a]
1	FZ		0.92	L	B
2	FZ		0.94	L	A
3	Epi	A	0.94	L	B
4	FZ		0.94	H	B
5	FZ		0.96	H	B
6	Epi		0.92	L	B
7	FZ		0.94	L	A
8	FZ	B	0.94	L	B
9	FZ		0.96	L	B
10	FZ		0.96	H	B
11	FZ		0.92	L	B
12	EPI		0.94	L	B
13	FZ	C	0.94	L	B
14	EPI		0.96	H	B
15	FZ		0.96	H	C

[a] In arbitrary unit.

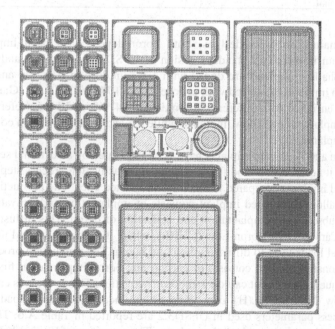

Figure A.5 RSD1 reticle layout.

Table A.6
Wafers of the UFSD3.2 Production

Wafer #	thickness [μm]	GI depth	p-gain dose[a]	C-dose[a]	diffusion process	V_{GL} [V]
1	45		1.00	1		23±0.5
2	45		1.00	1		23±0.5
3	45		1.00	0.8		23.5±0.5
4	45	shallow	1.00	04	CHBL	24.75±0.5
5	25		0.94	1		22±0.5
6	35		0.94	1		22±0.5
7	55		1.00	1		22.5±0.5
8	45		0.70	1		37.5±0.5
9	55		0.70	1		36.5±0.5
10	45	deep	0.70	0.6	CBL	41.5±0.5
11	45		0.70	-		47.75±0.5
12	45		0.74	1		39.75±0.5
13	45		0.74	0.6		44.5±0.5
14	45		0.74	1		45.3±0.5
15	55		0.74	1		44.5±0.5
16	45	deep	0.74	0.6	CBH	48.5±0.5
17	45		0.74	-		57.8±0.5
18	45		0.78	1		47.5±0.5
19	45		0.78	0.6		53.5±0.5

[a] In arbitrary unit.

Two main gain layer designs, both boron-doped, have been implemented: shallow-gain implant (the standard design of UFSD2 and UFSD3) and deep-gain implant. Shallow implants are doped with a p-gain dose of 0.98 a.u. and 0.94 a.u. while deep implants are doped with p-doses 0.70, 0.74, and 0.78 a.u.. Gain implants have been enriched with different carbon doses: 0.4, 0.8, and 1 a.u. (reference) have been co-implanted in shallow implants, while 0.6 and 1 a.u. have been co-implanted in deep implants.

Carbon and deep gain implants have been implanted by an external service. This fact made it necessary to modify the implantation-activation process steps for deep-carbonated implants compared to shallow-carbonated ones. The production process for the shallow-carbonated implant consists of implantation and activation of carbon and subsequently implantation and activation of boron. This diffusion process is called Carbon-High Boron-Low (CHBL), where the labels high and low indicate the thermal loads used to diffuse carbon and boron. The production process for the deep-carbonated implant consists of carbon and boron implantation first and then the subsequent concurrent carbon and boron activation. This process is called Caron Boron Low (High), CBL(H), depending on the thermal activation load used. The main process parameters used in UFSD3.2 are reported in Table A.6. The last column reports the gain layer depletion voltage. The extraction of V_{GL} has been obtained with a fit to the $I(V)$ characteristics, with an estimated accuracy of $\sigma(V_{GL}) = 0.5$ V.

Figure A.6 UFSD3.2 reticle layout.

In this production, the lithographic stepper technology has been used; Fig. A.6 shows the UFSD3.2 reticle layout. The reticle layout includes a PIN-LGAD couple (pad size of 1 mm^2), a single pad LGAD and 2×2 (5×5) array devices (pad size of 1.3×1.3 mm^2) with nine (three) different inter-pad distances. Table A.7 summarizes the inter-pad designs implemented in UFSD3.2. In UFSD3.2, a p-stop implant with a dose of 0.1 a.u., ten times lower than in UFSD2 and UFSD3, have been used.

A.2 CENTRO NACIONAL DE MICROELECTRONICA

A.2.1 RUN 12916 (CNM1)

The production comprises four high-resistivity p-type wafers, with an active thickness of 50 μm. All wafers have the same gain implant dose. The wafer layout contains single pads (1×1 mm^2 and 1.3×1.3 mm^2) and arrays with 2×2 (1×1 mm^2 and 1.3×1.3 mm^2) and 5×5 pads (1.3×1.3 mm^2). The layout is shown in Fig. A.7.

Three different inter-pad gaps, indicated as IP37, IP47, and IP57 have been implemented: IP37 is the shortest while IP57 the longest one.

Table A.7
UFSD3.2 Inter-pad Layouts in Array Devices

Device geometry	Inter-pad design	Inter-pad type	Inter-pad nominal distance [μm]
		1	16
	Single *p*-stop + extra ring	7	28
		8	28
2 × 2		2	21
	Single *p*-stop	11	21
		4	24
		5	25
	Double *p*-stop	9	38
	Double *p*-stop + bias grid	10	49
5 × 5	Single *p*-stop	8	28
	Double *p*-stop	9	38
	Double *p*-stop + bias grid	10	49

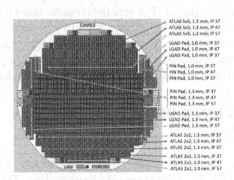

Figure A.7 CNM1 wafer layout [12].

Productions

167

A.3 HAMAMATSU PHOTONICS

A.3.1 ECX20840

HPK, in 2017, completed the ECX20840 run. This production comprises 16 high-resistivity p-type wafers, with active thicknesses of 50 and 80 μm. Four different p-gain doping concentrations have been implanted. The p-gain doses are reported in Table A.8. The label '1' corresponds to the lowest dose, '4' to the highest one. This production aimed to achieve very good temporal resolution. For this reason, very simple device geometries (single pad and arrays with 2×2 pads) have been implemented in the wafer layout.

Table A.8
Wafers of the ECX20840 Production

Wafer #	Bulk thickness [μm]	Split of p-gain dose
1-2		1
4-5	50	2
7-8		3
10-11		4
13-14		1
16-17	80	2
19-20		3
22-23		4

A.3.2 EXX28995 (HPK1)

EXX298995 (also called HPK1) was completed in 2018. This was the first production oriented to the R&D of the ATLAS and CMS timing detectors, aiming to demonstrate the feasibility of segmented large area UFSDs. HPK1 consists of 20 6″ p-type wafers with a high-resistivity bulk of active thickness of ~ 45 μm with a single p-gain dose implanted. Two designs of multiplication layers have been used: one shallower and wider (Type 3.1, wafer 1–10), and one deeper and narrower (Type 3.2, wafer 11–20).

The wafer layout of this production is equally divided into two main blocks, one for ATLAS and one for CMS, see Fig. A.8. The ATLAS devices consist of single pads, 2×2, 3×3, 5×5, and 15×15 arrays (pad size of 1.3×1.3 mm^2). Similarly, the CMS devices consist of single pads, 2×2, 3×3, 4×4, and 4×24 arrays (pad size of 1×3 mm^2).

Four different inter-pad gaps, indicated as 30, 50, 70, and 95, have been implemented: 30 indicates the shortest inter-pad, while 95 the longest.

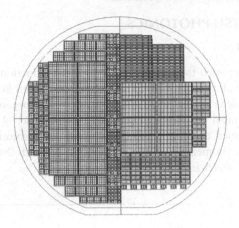

Figure A.8 HPK1 wafer layout [110].

A.3.3 EXX30327-EXX30328-EDX30329

In 2019, HPK produced three small batches of UFSDs (15 wafers in total), with an active thickness of ~ 35 μm and with the same wafer layout of the production EXX28995. In these batches, three different gain layers and bulk resistivity have been implemented, see Table A.9. Sensors called Type 1.1 have deep gain implants and a low-resistivity bulk; Type 1.2 has a deep gain implant in a high-resistivity bulk, while Type 2 a broad gain implant in a low-resistivity bulk.

Table A.9
EXX30327-EXX30328-EDX30329 Productions

Batch	Gain layer type	GI depth	Bulk resistivity [Ωcm]
EXX30327	1.1	Deep	Low
EXX30328	1.2	Deep	High
EDX30329	2	Broad	Low

A.3.4 HPK2

HPK2 is the second sensor HPK production focused on the R&D for the ATLAS and CMS timing detectors. This production has a deep and narrow multiplication layer, implanted in a high-resistivity bulk. Four splits of p-gain dose (1, 2, 3, and 4) have been implanted; split one corresponds to the highest p-dose, split 4 to the lowest. The targeted breakdown voltage at room temperature ranges from 160 V (split 1) to 240 V (split 4).

HPK2 comprises 16 wafers and two wafer layouts, shown in Fig. A.9 and listed in Table A.10; eight wafers contains smaller devices and the other eight larger ones. The *small sensors* layout includes devices with single pad and arrays of 2×2, 3×3, and 5×5 pads (pad-size of 1.3×1.3 mm^2). The *large sensors* layout includes arrays of 5×5, 8×8, 15×15, 16×16, 30×15, and 32×16 pads (pad-size of 1.3×1.3 mm^2).

Table A.10
Wafers of the HPK2 Production

Wafer #	Wafer layout	Split of p-gain dose	Target breakdown voltage [V]
25, 28	Small	1	160
31, 33	Small	2	180
36, 37	Small	3	220
42, 43	Small	4	240
1, 2	Large	1	160
7, 8	Large	2	180
14, 16	Large	3	220
19, 21	Large	4	240

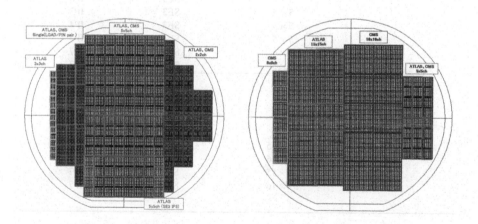

Figure A.9 HPK2 wafer layouts. *Left*: small sensors. *Right*: large sensors [110].

Four inter-pads layouts, labelled IP3 (shortest), IP4, IP5, and IP7, and two distances between the gain region and the edge of the sensor have been implemented, SE3 (300 μm) and SE5 (500 μm) , see Table A.11.

Table A.11
Devices in the HPK2 Production

Layout	Device geometry	Edge	Inter-pad
	Single pad	SE3	-
	2 × 2	SE3	IP3
	2 × 2	SE3	IP4
	2 × 2	SE3	IP5
	2 × 2	SE3	IP7
Small	2 × 2	SE5	IP5
	3 × 3	SE3	IP5
	5 × 5	SE3	IP3
	5 × 5	SE3	IP4
	5 × 5	SE3	IP5
	5 × 5	SE3	IP7
	5 × 5	SE5	IP7
	5 × 5	SE3	IP4
	5 × 5	SE3	IP5
	5 × 5	SE3	IP7
	5 × 5	SE5	IP7
Large	8 × 8	SE5	IP7
	15 × 15	SE3	IP7
	16 × 16	SE5	IP7
	30 × 15	SE3	IP7
	32 × 16	SE5	IP7

Acronyms

APD: Avalanche Photodiodes
BL, BH: boron low, high diffusion
BTBT: Band-To-Band Tunneling
BC: Boundary Conditions
BTE: Boltzmann Transport Equation
CCD: Charge-Coupled Device
CCE: Charge Collection Efficiency
CFD: Constant Fraction Discriminator
CNM: Centro Nacional de Microelectronica
DD: Drift-Diffusion
Deep-BL(H): Deep Boron-doped Low(High) Diffusion
DUT: Device Under Test
EM: Electromagnetic model
FB: Finite Boxes
FBK: Fondazione Bruno Kessler
FE: Finite Element
FD: Finite Differences
FZ: Float Zone
GR: Generation-Recombination
JTE: Junction Termination Extension
HPK: Hamamatsu Photonics
HV-SMU: High Voltage Source Monitor Unit
IC: Initial Conditions
IR: Infrared
MF-CMU: Multi-Frequency Capacitance Measurement Unit
MIP: Minimum Ionizing Particle
MP-SMU: Medium Power Source Monitor Unit
MPV: Most Probable Value
NIEL: Non Ionizing Energy Loss
LGAD: Low-Gain Avalanche Diode
PCB: Printed Circuit Board
RSD: Resistive AC-Coupled Silicon Detector
SG: Scharfetter-Gummel
SIMS: Secondary Ion Mass Spectrometer
SiPM: Silicon Photon Multiplier
SRH: Shockley-Read-Hall
Shallow-BL(H): Shallow Boron-doped Low(High) Diffusion
TAT: Trap-Assisted Tunneling
TCAD: Technology Computer-Aided Design
TCT: Transient Current Technique

TDC: Time-to-Digital Converter
TI: Trench Isolation
TM: Transport Model
ToT: Time-over-Threshold
UFSD: Ultra-Fast Silicon Detector
WF2: Weightfield2

References

1. Snowmass 2021. https://snowmass21.org/.

2. H. Bethe and W. Heitler. On the stopping of fast particles and on the creation of positive electrons. *Proc. Royal Soc. London*, 146(856):83, 1934.

3. D.M. Caughey and R.E. Thomas. Carrier mobilities in silicon empirically related to doping and field. *Proc. IEEE*, 55(12):2192–2193, 1967.

4. C.E. Jones et al. Carbon-acceptor pair centers (X centers) in silicon. *J. Appl. Phys.*, 52:5148, 1981.

5. F. Cenna. *Design and Test of Sensors and Front-End Electronics for Fast Timing in High Energy Physics*. PhD thesis, Università degli Studi di Torino, 2018.

6. Centro Nacional de Microelectronica. http://www.imb-cnm.csic.es/index.php/en/.

7. A.G. Chynoweth. Ionization rates for electrons and holes in Silicon. *Phys. Rev.*, 109(5):1537, 1958.

8. ATLAS Collaboration. Technical Proposal: A High-Granularity Timing Detector for the ATLAS Phase-II Upgrade. *CERN-LHCC-2018-023, LHCC-P-012*, 2018.

9. CMS Collaboration. A MIP Timing Detector for the CMS Phase-2 Upgrade. *Technical Design Report, CERN-LHCC-2019-003, CMS-TDR-020*, 2019.

10. RD53 Collaboration. RD Collaboration Proposal: Development of pixel readout integrated circuit for extreme rate and radiation. *CERN-LHCC-2013-008, LHCC-P-006*, 2013.

11. The ROSE Collaboration. Third RD48 status report. *CERN-LHCC 2000-009, LEB Status Report RD48*, 2000.

12. Centro Nacional de Microelectronica. Private communication.

13. B. Delaunay. Sur la sphere vide. *Bull. Acad. Sci. USSR(VII), Classe Sci. Mat. Nat.*, 793, 1934.

14. A. Apresyan et al. Measurements of an AC-LGAD strip sensor with a 120 GeV proton beam. *J. Instrum.*, 15:P09038, 2020.

15. A. Howard et al. LGAD measurements from different producers. In *37th RD50 Workshop*, 2020.

16. A. Morozzi et al. TCAD advanced radiation damage modeling in silicon detectors. *PoS Proc. Sci.*, 373(Vertex2019):050, 2020.

17. A. Tournier et al. Pixel-to-Pixel isolation by Deep Trench technology: Application to CMOS Image Sensor. In *IISW 2011*, 2011.

18. A. Vignati et al. Innovative thin silicon detectors for monitoring of therapeutic proton beams: Preliminary beam tests. *J. Instrum.*, 12:C12056, 2017.

19. A. Vignati et al. A new detector for the beam energy measurement in proton therapy: A feasibility study. *Phys. Med. Biol.*, 65:215030, 2020.

20. A. Vignati et al. Thin low-gain avalanche detectors for particle therapy applications. *J. Phys. Conf. Ser.*, 1662:012035, 2020.

21. B. Gkotse et al. Irradiation Facilities at CERN. In *Radecs 2017*, 2017.

22. C. De La Taille et al. ALTIROC0, a 20 pico-second time resolution ASIC for the ATLAS High Granularity Timing Detector (HGTD). *PoS Proc. Sci.*, TWEPP-17:006, 2018.

23. C. Piemonte et al. Performance of NUV-HD Silicon Photomultiplier Technology. *IEEE T. Electron Dev.*, 63(3):1111, 2016.

24. E.J. Olave et al. FAST: A 30 ps time resolution front-end ASIC for a 4D tracking system based on Ultra Fast Silicon Detectors. *Nucl. Inst. Meth. A*, 985:164615, 2021.

25. F. Anghinolfi et al. NINO: An ultrafast low-power front-end amplifier discriminator for the time-of-flight detector in the ALICE experiment. *IEEE T. Nucl. Sci.*, 51:1974, 2004.

26. F. Cenna et al. Weightfield2: a fast simulator for silicon and diamond solid state detector. *Nucl. Inst. Meth. A*, 796:149, 2015.

27. F. Cenna et al. TOFFEE: a full custom amplifier-comparator chip for timing applications with silicon detectors. *J. Instrum.*, 12:C03031, 2017.

28. F. Moscatelli et al. Analysis of surface radiation damage effects at HL-LHC fluences: Comparison of different technology options. *Nucl. Inst. Meth. A*, 924:198, 2019.

29. A. Morozzi et al. TCAD Modeling of Surface Radiation Damage Effects: A State-Of-The-Art Review. *Front. Phys.*, 9:617322, 2021.

30. F. Siviero et al. Characterization with a β-source setup of the FBK UFSD 3.2 and HPK2 LGAD productions. In *37th RD50 Workshop*, 2020.

31. G-F. Dalla Betta et al. Design and TCAD simulation of double-sided pixelated low gain avalanche detectors. *Nucl. Inst. Meth. A*, 796:154, 2015.

32. G. Giacomini et al. Development of a technology for the fabrication of Low-Gain Avalanche Diodes at BNL. *Nucl. Inst. Meth. A*, 934:52, 2019.

33. G. Kramberger et al. Charge collection properties of heavily irradiated epitaxial silicon detectors. *Nucl. Inst. Meth. A*, 554:212, 2005.

34. G. Kramberger et al. Charge collection studies on custom silicon detectors irradiated up to $1.6 \cdot 10^{17}$ n_{eq}/cm^2. *J. Instrum.*, 8:P08004, 2013.

35. G. Kramberger et al. Radiation effects in Low Gain Avalanche Detectors after hadron irradiations. *J. Instrum.*, 10:P07006, 2015.

36. G. Kramberger et al. Radiation hardness of thin Low Gain Avalanche Detectors. *Nucl. Inst. Meth. A*, 891:68, 2018.

37. G. Kramberger et. al. Annealing effects on operation of thin Low Gain Avalanche Detectors. *Submitted to J. Instrum.*, 2020. arXiv: 2005.14556.

38. G. Paternoster et al. Developments and first measurements of Ultra-Fast Silicon Detectors produced at FBK. *J. Instrum.*, 12:C02077, 2017.

39. G. Paternoster et al. Trench-Isolated Low Gain Avalanche Diodes (TI-LGADs). *IEEE Electr. Device L.*, 41(6):884, 2020.

40. G. Pellegrini et al. Technology developments and first measurements of Low Gain Avalanche Detectors (LGAD) for high energy physics applications. *Nucl. Inst. Meth. A*, 765:12, 2014.

41. I. Mandić et al. Measurements with silicon detectors at extreme neutron fluences. *J. Instrum.*, 15:P11018, 2020.

42. I. Pintilie et al. Stable radiation-induced donor generation and its influence on the radiation tolerance of silicon diodes. *Nucl. Inst. Meth. A*, 556:197, 2006.

43. J. Lange et al. Gain and time resolution of 45 μm thin Low Gain Avalanche Detectors before and after irradiation up to a fluence of 10^{15} n_{eq}/cm^2. *J. Instrum.*, 12:P05003, 2017.

44. J. Lange et al. Radiation hardness of small-pitch 3D pixel sensors up to a fluence of $3 \cdot 10^{16}$ n_{eq}/cm^2. *J. Instrum.*, 13:P09009, 2018.

45. M. Andrä et al. Development of low-energy X-ray detectors using LGAD sensors. *J. Synchrotron Radiat.*, 26:1226, 2019.

46. M. Berretti et al. Test of Ultra Fast Silicon Detectors for the TOTEM upgrade project. *J. Instrum.*, 12:P03024, 2017.

47. M. Ferrero et al. Study and characterization of low gain avalanche diode. In *TREDI2017: 12th "Trento" Workshop on Advanced Silicon Radiation Detectors*, 2017.

48. M. Ferrero et al. Radiation resistant LGAD design. *Nucl. Inst. Meth. A*, 919:16, 2019.

49. M. Ferrero et al. Recent studies and characterization on UFSD sensors. In *34th RD50 Workshop*, 2019.

50. M. Ferrero et al. Evolution of the design of ultra fast silicon detector to cope with high irradiation fluences and fine segmentation. *J. Instrum.*, 15:C04027, 2020.

51. M. Mandurrino et al. Progresses in LGAD simulations and comparison with experimental data from UFSD2, the new 50-μm production at FBK. In *31st RD50 Workshop*, 2017.

52. M. Mandurrino et al. TCAD simulation of silicon detectors: A validation tool for the development of LGAD. In *30th RD50 Workshop*, 2017.

53. M. Mandurrino et al. Demonstration of 200-, 100-, and 50-μm Pitch Resistive AC-Coupled Silicon Detectors (RSD) With 100% Fill-Factor for 4D Particle Tracking. *IEEE Electr. Device L.*, 40(11):1780, 2019.

54. M. Mandurrino et al. Analysis and numerical design of Resistive AC-Coupled Silicon Detectors (RSD) for 4D particle tracking. *Nucl. Inst. Meth. A*, 959:163479, 2020.

55. M. Mandurrino et al. High performance picosecond- and micron-level 4D particle tracking with 100% fill-factor Resistive AC-Coupled Silicon Detectors (RSD), 2020.

56. M. Manna et al. Characterisation of 3D pixel sensors irradiated at extreme fluences. In *35th RD50 Workshop*, 2019.

57. M. Tornago et al. Performances of the third UFSD production at FBK. In *33rd RD50 Workshop*, 2018.

58. M. Tornago et. al. Resistive AC-Coupled Silicon Detectors: Principles of operation and first results from a combined analysis of beam test and laser data. *Submitted to Nucl. Inst. Meth. A*, 2020. arXiv: 2007.09528.

59. M.K. Petterson et al. Charge collection and capacitance–voltage analysis in irradiated n-type magnetic Czochralski silicon detectors. *Nucl. Inst. Meth. A*, 583:189, 2007.

60. N. Cartiglia et al. Design optimization of ultra-fast silicon detectors. *Nucl. Inst. Meth. A*, 796:141, 2015.

61. N. Cartiglia et al. Issues in the design of Ultrafast silicon detectors. In *TREDI2015: 10th "Trento" Workshop on Advanced Silicon Radiation Detectors*, 2015.

62. N. Cartiglia et al. The 4D pixel challenge. *J. Instrum.*, 11:C12016, 2016.

63. N. Cartiglia et al. Beam test results of a 16ps timing system based on ultra-fast silicon detectors. *Nucl. Inst. Meth. A*, 850:83, 2017.

64. N. Cartiglia et al. Tracking in 4 dimensions. *Nucl. Inst. Meth. A*, 845:47, 2017.

65. N. Cartiglia et al. Tracking in 4 dimensions. *PoS Proc. Sci.*, 314(EPS-HEP2017):489, 2017.

66. N. Cartiglia et al. LGAD designs for Future Particle Trackers. *Nucl. Inst. Meth. A*, 979:164383, 2020.

67. P.A. Zyla et al. Review of Particle Physics. *Prog. Theor. Exp. Phys.*, 2020(8):083C01, 2020.

68. R. Arcidiacono et al. State-of-the-art and evolution of UFSD sensors design at FBK. *Nucl. Inst. Meth. A*, 978:164375, 2020.

69. R. Sacchi et al. Test of innovative silicon detectors for the monitoring of a therapeutic proton beam. *J. Phys. Conf. Ser.*, 1662:012002, 2020.

70. S. Reggiani et al. Measurement and modeling of the electron impact-ionization coefficient in silicon up to very high temperatures. *IEEE T. Electron Dev.*, 52(10):2290, 2005.

71. S. Terzo et al. Heavily irradiated N-in-p thin planar pixel sensors with and without active edges. *J. Instrum.*, 9:C05023, 2014.

72. S.M. Mazza et al. Properties of FBK UFSDs after neutron and proton irradiation up to $6 \cdot 10^{15}$ n_{eq}/cm^2. *J. Instrum.*, 15:T04008, 2020.

73. T. Liu et al. "The ETROC Project: Precision Timing ASIC Development for LGAD-based CMS Endcap Timing Layer (ETL) Upgrade". In *TWEPP 2019 Topical Workshop on Electronics for Particle Physics*, 2019.

74. V. Eremin et al. Development of transient current and charge techniques for the measurement of effective net concentration of ionized charges (N_{eff}) in the space charge region of *p-n* junction detectors. *Nucl. Inst. Meth. A*, 372(3):388–398, 1993.

75. V. Sola et al. Ultra-Fast Silicon Detectors for 4D tracking. *J. Instrum.*, 12:C02072, 2017.

76. V. Sola et al. First FBK production of 50 μm ultra-fast silicon detectors. *Nucl. Inst. Meth. A*, 924:360, 2019.

77. V. Sola et al. First results from thin silicon sensors for extreme fluences . In *37th RD50 Workshop*, 2020.

78. V. Sola et al. Next-Generation Tracking System for Future Hadron Colliders. *PoS Proc. Sci.*, 373(Vertex2019):034, 2020.

79. X. Yu et al. Effect of germanium on the kinetics of boron-oxygen defect generation and dissociation in Czochralski silicon. *Appl. Phys. Lett.*, 97:162107, 2010.

80. X. Yu et al. Suppression of boron–oxygen defects in p-type Czochralski silicon by germanium doping. *Appl. Phys. Lett.*, 97:139901, 2010.

81. Y. Unno et al. Optimization of surface structures in n-in-p silicon sensors using TCAD simulation. *Nucl. Inst. Meth. A*, 636:S118, 2011.

82. Z. Galloway et al. Properties of HPK UFSD after neutron irradiation up to 6e15 n/cm². *Nucl. Inst. Meth. A*, 940:19, 2019.

83. European Strategy for Particle Physics. https://europeanstrategy.cern/.

84. M. Friedl. *The CMS Silicon Strip Tracker and its Electronic Readout*. PhD thesis, Vienna University of Technology, 2001.

85. M. Garcia-Sciveres and N. Wermes. A review of advances in pixel detectors for experiments with high rate and radiation. *Rep. Prog. Phys.*, 81:066101, 2018.

86. F. Hartmann. *Evolution of Silicon Sensor Technology in Particle Physics*. Springer, 2009.

87. M. Huhtinen. Simulation of non-ionising energy loss and defect formation in silicon. *Nucl. Inst. Meth. A*, 491:194, 2002.

88. G.A.M. Hurkx. On the modelling of tunnelling currents in reverse-biased *p-n* junctions. *Solid-State Electron.*, 32(8):665, 1989.

89. G.A.M. Hurkx, D.B.M. Klaassen, and M.P.G. Knuvers. A new recombination model for device simulation including tunneling. *IEEE T. Electron Dev.*, 39(2):331, 1992.

90. CYRIC Proton Irradiation. https://www.cyric.tohoku.ac.jp/english/index.html.

91. KIT Proton Irradiation. https://www.etp.kit.edu/english/264.php.

92. C. Jacoboni, C. Canali, G. Ottaviani, and A. Alberigi Quaranta. A review of some charge transport properties of silicon. *Solid-State Electron.*, 20(2):77, 1977.

93. E.O. Kane. Zener tunneling in semiconductors. *J. Phys. Chem. Solids*, 12(2):181, 1960.

94. Fondazione Bruno Kessler. https://www.fbk.eu/en/.

95. D.B.M. Klaassen, J.W. Slotboom, and H.C. de Graaff. Unified apparent bandgap narrowing in *n*- and *p*-type Silicon. *Solid-State Electron.*, 35(2):125, 1992.

96. G. Kramberger. Reasons for high charge collection efficiency of silicon detectors at HL-LHC fluences. *Nucl. Inst. Meth. A*, 924:192, 2019.

97. G. Lutz. *Semiconductor radiation detectors: device physics*. Springer, 1999.

98. W. Maes, K. De Meyer, and R. Van Overstraeten. Impact ionization in silicon: A review and update. *Solid State Electron.*, 33:705, 1999.

99. M. Mandurrino. *Advances in Quantum Tunneling Models for Semiconductor Optoelectronic Device Simulation*. PhD thesis, Politecnico di Torino, 2017.

100. D.J. Massey, J.P.R. David, and G.J. Rees. Temperature dependence of impact ionization in submicrometer Silicon devices. *IEEE T. Electron Dev.*, 53(9):2328, 2006.

101. S. Meroli, D. Passeri, and L. Servoli. Energy loss measurement for charged particles in very thin silicon layers. *J. Instrum.*, 6:P06013, 2011.

102. M. Moll. Displacement Damage in Silicon Detectors for High Energy Physics. *IEEE T. Nucl. Sci.*, 65(8):1561, 2018.

103. M. Moll. Acceptor removal - Displacement damage effects involving the shallow acceptor doping of p-type silicon devices. *PoS Proc. Sci.*, 373(Vertex2019):027, 2020.

104. M. Mota and J. Christiansen. A high-resolution time interpolator based on a delay locked loop and an RC delay line. *IEEE J. Solid-St. Circ.*, 34(10):1360, 1999.

105. Y. Okuto and C.R. Crowell. Threshold energy effect on avalanche breakdown voltage in semiconductor junctions. *Solid-State Electron.*, 18(2):161, 1975.

106. R. Arcidiacono on behalf of the CMS and TOTEM collaborations. A new timing detector for the CT-PPS project. *Nucl. Inst. Meth. A*, 845:16, 2017.

107. S.I. Parker, C.J. Kenney, and J. Segal. 3D – A proposed new architecture for solid-state radiation detectors. *Nucl. Inst. Meth. A*, 395:328, 1997.

108. Particulars. http://particulars.si.

109. Hamamatsu Photonics. https://www.hamamatsu.com/eu/en/index.html.

110. Hamamatsu Photonics. Private communication.

111. Los Alamos Neutron Science Center (LANSCE) proton accelerator. https://lansce.lanl.gov.

112. S. Ramo. Currents Induced by Electron Motion. *IEEE Proceedings of the IRE*, 27(9):584, 1939.

113. Radiation hard semiconductor devices for very high luminosity colliders RD50 Collaboration. http://rd50.web.cern.ch/rd50/.

114. A. Rivetti. Fast front-end electronics for semiconductor tracking detectors: Trends and perspectives. *Nucl. Inst. Meth. A*, 765:202, 2014.

115. A. Rivetti. *CMOS: Front-end Electronics for Radiation Sensors*. CRC Press, 2015.

116. L. Rossi, P. Fisher, T. Rohe, and N. Wermes. *Pixel Detectors*. Springer, 2006.

117. H. F.-W. Sadrozinski. Timing measurements on Ultra-Fast Silicon Detectors. In *TREDI2017: 12th "Trento" Workshop on Advanced Silicon Radiation Detector*, 2017.

118. H.F.-W. Sadrozinski, A. Seiden, and N. Cartiglia. 4D tracking with ultra-fast silicon detectors. *Rep. Prog. Phys.*, 81:026101, 2017.

119. W. Shockley. Current to Conductors Induced by a Moving Point Charge. *J. Appl. Phys.*, 9:635, 1938.

120. W. Shockley and W.T. Read, Jr. Statistics of the recombinations of holes and electrons. *Phys. Rev.*, 87(5):835, 1952.

121. F. Siviero. Development of ultra-fast silicon detectors for time measurements at High-Luminosity LHC. Master's thesis, Università degli Studi di Torino, 2018.

122. F. Siviero. Characterization of the FBK UFSD3.2 production using a β source. In *TREDI2020: 16th "Trento" Workshop on Advanced Silicon Radiation Detectors*, 2021.

123. J.W. Slotboom. The *pn*-product in silicon. *Solid-State Electron.*, 20(4):279, 1977.

124. J.W. Slotboom and H.C. de Graaff. Measurements of bandgap narrowing in Si bipolar transistors. *Solid-State Electron.*, 19(10):857, 1976.

125. J.W. Slotboom and H.C. de Graaff. Bandgap narrowing in silicon bipolar transistors. *IEEE T. Electron Dev.*, 24(8):1123, 1977.

126. L. Snoj, G. Žerovnik, and A. Trkov. Computational analysis of irradiation facilities at the JSI TRIGA reactor. *Appl. Radiat. Isot.*, 70(3):483, 2012.

127. V. Sola. Silicon Sensors for Extreme Fluences. In *TREDI2020: 15th "Trento" Workshop on Advanced Silicon Radiation Detectors*, 2020.

128. Synopsys. *Sentaurus Device User Guide*.

129. S.M. Sze. *Physics of Semiconductor Devices*. John Wiley and Sons, 2015.

130. T. Croci et al. TCAD numerical simulation of irradiated Low-Gain Avalanche Diodes. In TREDI2021: 16th "Trento" Workshop on Advanced Silicon Radiation Detectors, 2021.

131. M. Tornago. Development of Ultra-Fast Silicon Detectors for 4D tracking at High Luminosity LHC: laboratory measurements and numerical simulations. Master's thesis, Università degli Studi di Torino, 2019.

132. J. Vaitkus. An increase of the quantum yield in highly irradiated Si. In *37th RD50 Workshop*, 2020.

133. R. van Overstraeten and H. de Man. Measurement of the ionization rates in diffused Silicon *p-n* junctions. *Solid-State Electron.*, 13(5):583, 1970.

134. A. Vasilescu and G. Lindström. Displacement damage in silicon, online compilation, 2000. https://rd50.web.cern.ch/NIEL/.

Index

Printed in the United States
by Baker & Taylor Publisher Services

Printed in the United States
by Baker & Taylor Publisher Services